Fundamentals of Electromechanical Drives

Online at: https://doi.org/10.1088/978-0-7503-6104-0

ISBN   978-0-7503-6104-0 (ebook)
ISBN   978-0-7503-6102-6 (print)
ISBN   978-0-7503-6105-7 (myPrint)
ISBN   978-0-7503-6103-3 (mobi)

DOI   10.1088/978-0-7503-6104-0

Version: 20250201

IOP ebooks

British Library Cataloguing-in-Publication Data: A catalogue record for this book is available from the British Library.

Published by IOP Publishing, wholly owned by The Institute of Physics, London

IOP Publishing, No.2 The Distillery, Glassfields, Avon Street, Bristol, BS2 0GR, UK

US Office: IOP Publishing, Inc., 190 North Independence Mall West, Suite 601, Philadelphia, PA 19106, USA

In memory of my beloved wife Lea and our sweet daughter Maya.

Contents

Preface

The function of an electromechanical drive system is to transmit motion to a working machine. It consists of equipment designed for converting electrical to mechanical power suitable for operating the machine. Several examples of drive systems are: elevators, cranes, hoists, conveyers, machine tools, pumps, fans, robots, and electrical traction systems.

The design of an electromechanical drive system involves consideration of many factors such as steady-state performance, starting, dynamic and regenerative braking, plugging and reverse direction operation, speed control, sudden and temporary overloads, ambient conditions and mechanical coupling. This textbook treats the above topics through theory and analytical studies that include technical understanding of mechanical characteristics of the electrical motors as well as the driven machines. The book material, the order of the chapters, and the drilling problems are based on years of teaching experience including students' reaction during and after class and their knowledge acquisition.

The content of this textbook addresses the fundamentals of electromechanical drive systems and it covers translation of load performance characteristics to the motor shaft; electromechanical energy conversion; acceleration and deceleration time; construction of load diagram; choice of motor type and size for different duty cycles; four quadrant motor operation; basics of DC (direct current), induction, synchronous, and brushless DC motors; electrical braking modes; conventional and modern speed control of DC and AC drives. Also included are many worked examples taken from practical electromechanical drive systems.

The material of this textbook is intended first as a beginning graduate level course, The entire text can be covered in one academic semester (15 weeks). A prerequisite course on the fundamental of electric power systems analysis is highly recommended. Although reviews of technical material in appropriate chapters are included in the book, an advance knowledge of rotating machines and power electronics would be helpful to the reader. Second, the book material is also relevant to design and supervising engineers working in the area of electrical power systems for utility, architects and engineering firms, and commercial and industrial companies. Various examples throughout the book with their detailed solutions and drilling problems included in different chapters would provide a better grasp of the analysis and the performance features of the major components of electromechanical drive systems. A solution manual for all drill problems is available for instructors (https://iopscience.iop.org/book/mono/978-0-7503-6104-0).

The book is divided into three main parts:

First part: The mechanical part, also called the working machine or simply the mechanism, is a general term given to such equipment as machine tools, hoists, winches, cranes, conveyers, elevators, pumps, tool machines, fans, and electrical traction systems. To adapt an electric motor to the working machine specifications, all mechanical motions must be referred to the motor shaft. This part of the book addresses methods of referring steady state and dynamic motions to one shaft and provides an analysis of the transient conditions of the system dynamics.

Second part: The electric motor is the prime mover of the electromechanical drive system. The motor is required to address the needs of the working machine in steady state operations, in controlling the mechanism speed, and in braking modes of operation. This part addresses three motor types: the shunt-wound connected direct current DC motor, the induction motor, and the synchronous motor operating as a brushless DC motor.

Third part: The correct catalogue selection of an electric motor that drives a mechanism is a matter of economic interest because it affects the capital and the running cost of the drive system, bearing in mind the enormous number of machines that employ electric motors throughout the industry. This part addresses major considerations for selecting an electric motor for specific working conditions, and the adaptation of engineering design solutions to standard industrial motor catalogues.

Acknowledgements

Through years of developing and teaching the course of electromechanical systems, I have received valuable help from many colleagues and would like to sincerely thank them all. A special thanks to Professor Abraham Alexandrovitz of the Technion, Israel Institute of Technology, Haifa, for his invaluable inspiration, pedagogy and guidance in the preparation for the manuscript for this textbook.

I also would like to thank IOP Publishing for their excellent work, especially Ms Caroline Mitchell, Isabelle Defillion, Betty Barber, and Mia Foulkes for their valuable support and help.

Two computer programs were extremely helpful in preparing the material for publication: LiveMath Maker, which is a computer algebra and graphing system, and Canvas X Draw, which is a graphic design software.

Author biography

Zivan Zabar

Zivan Zabar is Professor Emeritus of Electrical and Computer Engineering, NYU. He retired from Tandon School of Engineering on September 2020 after 45 years of service. He received his PhD from the Technion—Israel Institute of Technology in 1972. His areas of expertise are electric power systems, electric drives, and power electronics. He served as the Head of ECE Department for three years (1995–1998). Since 1979, he had research support from the US Government, NY State, as well as continuous research from the two electric utilities in the Metropolitan area, Consolidated Edison of NYC and National Grid of Long Island, NY. He received the Lady Davis Fellowship from the EE Department at the Technion, Israel in the Spring of 2000. He has six patents, one textbook, and more than 50 papers published in technical journals. He is a senior member of the IEEE, and a member of Sigma Xi.

IOP Publishing

Fundamentals of Electromechanical Drives

Zivan Zabar

Chapter 1

Introduction—basic concepts

The function of an electromechanical drive system is to transmit motion to a working machine. The drive system consists of equipment designed for converting electrical to mechanical power suitable for operating the machine. Several examples of drive systems are: elevators, cranes, hoists, conveyers, machine tools, pumps, fans, robots, and electrical traction systems. The design of a drive system involves consideration of many factors such as the steady state and transient performance, braking modes of operation, speed control, sudden and temporary overloads, ambient conditions, and mechanical couplings. This chapter provides an overview of the main system components, passive and dynamic mechanical motions, and the integration of the motor features and the mechanism specifications.

1.1 Main system components

An electromechanical drive is a form of machine equipment designed to convert electric energy into mechanical energy and provides control of that process [1–5]. The main function of the drive is the delivery of the motion tasks required by the mechanism, which includes its speed–torque characteristics through all feasible transient conditions. A fundamental block diagram of an electromechanical drive is shown in figure 1.1.

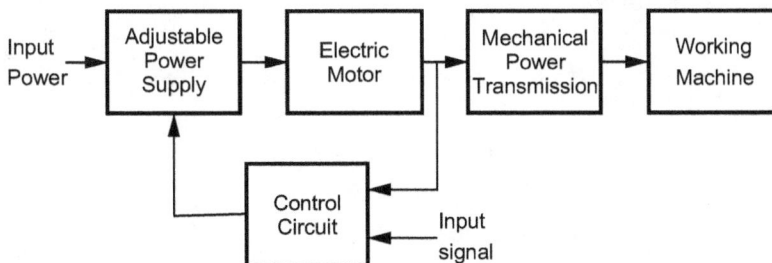

Figure 1.1. Fundamental block diagram of electromechanical drive.

doi:10.1088/978-0-7503-6104-0ch1   1-1

The drive consists of five basic elements (figure 1.1):
   (a) An adjustable power supply that controls the electric motor operation.
   (b) An electric motor that converts electrical to mechanical power.
   (c) A mechanical transmission to transfer torque and speed to the mechanism.
   (d) A mechanism (working machine) that performs the required mechanical work.
   (e) A control circuit.

The adjustable power supply unit might include switches, relays, transformers, and power electronic converters. The electric motor unit might include direct-current (DC), induction, and synchronous rotating machines. The mechanical power transmission unit might include gears, belts and pulleys, chainwheels, and power screws. The mechanism (working machine) unit might include a few moving parts working together to carry out the mechanical task. The control unit governs the adjustable power supply to perform accurately the required work.

1.2 Power, energy, mass, moment of inertia, and flywheel moment

This subchapter provides a review of relevant fundamental definitions and formulas that are used in following chapters.

1.2.1 Linear mechanical motion

In a linear moving mechanical system, the fundamental equation of motion suggests that the driving force is counter-balanced by the resistive force set up by the driven load and by the dynamic force arising from a change in velocity.

$$\left. \begin{aligned} F_d &= F_r + F_{dyn} = F_r + M\frac{dv}{dt} \ [\text{N}] \\ F_{dyn} &= M\frac{dv}{dt} \ [\text{N}] \end{aligned} \right\}$$
(1.1)

where
   F_d [N] is the driving force
   F_r [N] is the resistive force developed by the load
   F_{dyn} [N] is the dynamic force
   M [$\text{kg}_\text{m} \equiv \text{N} \cdot \text{s}^2 \ \text{m}^{-1}$] is the mass
   v [m s^{-1}] is the linear velocity, and
   t [s] is the time.

Unit of kinetic energy is:

$$dE = F \cdot dl \ [\text{joule}]$$

where
   E [joule $\equiv$ Nm $\equiv$ W $\cdot$ s] is energy
   F [N] is the force, and
   l [m] is the distance.

Power in [$W \equiv Nm\ s^{-1}$] is defined as change of energy per unit time:

$$P = \frac{dE}{dt} = F\frac{dl}{dt} = F \cdot v \ \ [W] \tag{1.2}$$

From equations (1.1) and (1.2), the kinetic energy stored in a linear moving mass is:

$$E = \int (F \cdot v)\, dt = \int M \cdot v \cdot dv = \frac{1}{2}M \cdot v^2 \ \ [joule] \tag{1.3}$$

1.2.2 Rotational mechanical motion

In a rotating mechanical system, the fundamental equation of motion suggests that the motor driving torque is counter-balanced by the resistive torque set up by the driven mechanism and by the dynamic torque arising from a change in shaft speed:

$$\left.\begin{array}{c} T_m = T_r + T_{dyn} = T_r + J\dfrac{d\omega}{dt} \ \ [Nm] \\[2ex] T_{dyn} = J\dfrac{d\omega}{dt} \ \ [Nm] \end{array}\right\} \tag{1.4}$$

where
T_m [Nm] is the driving torque
T_r [Nm] is the resistive torque developed by the mechanism
T_{dyn} [Nm] is the dynamic torque
J [$kg_m \cdot m^2 \equiv Nm \cdot s^2$] is the moment of inertia
ω [s^{-1}] is the angular velocity, and
t [s] is the time.

The angular velocity can be expressed in terms of revolutions per minute:

$$\omega = \frac{2\pi \cdot n}{60} \ \ [s^{-1}] \tag{1.5}$$

where n is the technical term for speed in revolutions per minute [rpm].
    The moment of inertia is defined as:

$$J = M \cdot \rho^2 \ \ [kg_m \cdot m^2 \equiv Nm \cdot s^2] \tag{1.6}$$

where ρ (sometimes denotes as k) is defined as the radius of gyration (see example 1.1 below).
    Unit of kinetic energy is:

$$dE = T \cdot d\theta \ \ [joule]$$

where
    E [joule $\equiv$ Nm $\equiv$ W $\cdot$ s] is energy

T [Nm $\equiv$ joule] is the torque, and
θ [Radian] is the angel.

Power is defined as:

$$P = \frac{dE}{dt} = T\frac{d\theta}{dt} = T \cdot \omega \ [\text{W} \equiv \text{Nm s}^{-1}] \tag{1.7}$$

From equations (1.4) and (1.7), the kinetic energy stored in a rotating object is:

$$E = \int (T \cdot \omega)dt = \int J \cdot \omega \cdot d\omega = \frac{1}{2}J \cdot \omega^2 \ [\text{joule}] \tag{1.8}$$

Example 1.1 Find the moment of inertia J and the radius of gyration ρ of a solid circular cylinder body of radius R and length L rotating at an angular speed ω around its central axis (figure E1.1).

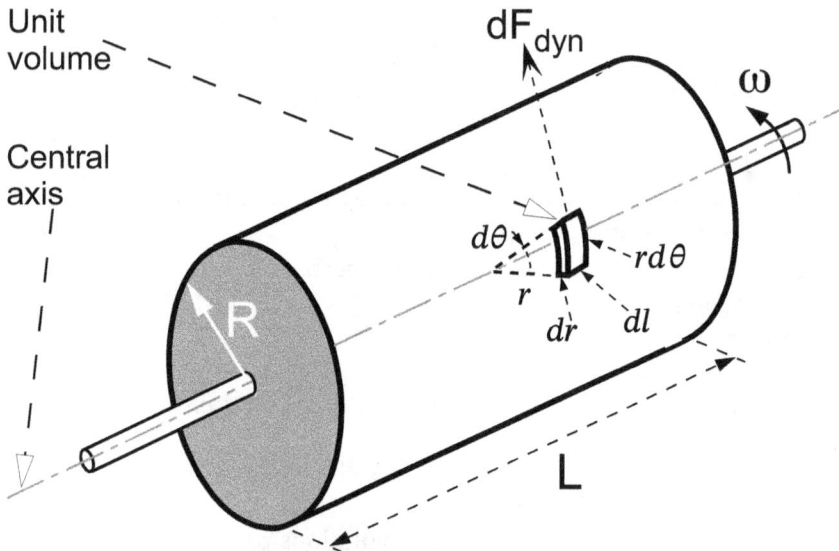

Figure E1.1. Solid circular cylinder rotating around its central axis.

Solution
The cylinder volume is:

$$V = \pi R^2 \cdot L \ [\text{m}^3]$$

An infinitesimal unit volume at a radius r from the central axis is shown in figure E1.1. That unit volume is:

$$dV = dr \cdot dl \cdot r \ d\theta \ [\text{m}^3]$$

A unit mass would be:

$$dM = \frac{M}{V}dV = \frac{M}{\pi R^2 \cdot L}dV \ [\mathrm{N} \cdot \mathrm{s}^2 \mathrm{m}^{-1}]$$

The tangential velocity of the unit volume is:

$$v = \omega \cdot r \ [\mathrm{ms}^{-1}]$$

Assume that the velocity v is constant, but its direction is continually changing. Hence, the body is considered to be in a constant angular acceleration. This angular acceleration requires a unit dynamic force (equation (1.1)):

$$dF_{\mathrm{dyn}} = dM\frac{dv}{dt} = dM\frac{d}{dt}(\omega \cdot r) = dM \cdot r\frac{d\omega}{dt} = dM \cdot r \cdot \alpha \ [\mathrm{N}]$$

where $\alpha = \frac{d\omega}{dt}$ [s^{-2}] is the angular acceleration

A unit dynamic torque dT_{dyn} is:

$$dT_{\mathrm{dyn}} = r \cdot dF_{\mathrm{dyn}} = r \cdot dM \cdot r \cdot \alpha \ [\mathrm{Nm}]$$

Substituting the unit mass dM into the above equation for dT_{dyn} suggests:

$$dT_{\mathrm{dyn}} = r^2 \cdot \frac{M}{\pi R^2 \cdot L}dV \cdot \alpha = \frac{M \cdot \alpha}{\pi R^2 \cdot L}r^3 \cdot dr \cdot dl \cdot d\theta \ [\mathrm{Nm}]$$

And the dynamic torque itself would be:

$$T_{\mathrm{dyn}} = \frac{M \cdot \alpha}{\pi R^2 \cdot L}\int_0^R \int_0^L \int_0^{2\pi} r^3 \cdot dr \cdot dl \cdot d\theta = M\frac{R^2}{2} \cdot \alpha = J\frac{d\omega}{dt} \ [\mathrm{Nm}]$$

where $J = M\dfrac{R^2}{2}$ [kg$_{\mathrm{m}}$ · m^2] is the moment of inertia

Using the definition for J (equation (1.6)):

$$\left\{ \begin{array}{l} J = M \cdot \rho^2 = M\dfrac{R^2}{2} \ [\mathrm{kg_m} \cdot \mathrm{m}^2 \equiv \mathrm{Nm} \cdot \mathrm{s}^2] \\[4mm] \rho = \dfrac{R}{\sqrt{2}} \ [\mathrm{m}] \end{array} \right\}$$

where ρ [m] is the radius of gyration of the solid circular cylinder when it rotates around its central axis.

Note: Reference [6] provides moments of inertia of additional bodies.

1.2.3 Flywheel moment

The technical term used in industry instead of moment of Inertia is called flywheel moment or flywheel effect. The relationship between the two terms is:

$$J = M \cdot \rho^2 = \frac{G}{g} \cdot \left(\frac{D}{2}\right)^2 = \frac{GD^2}{4g} \ [\mathrm{kg_m} \cdot \mathrm{m}^2] \tag{1.9}$$

and the flywheel moment is:

$$GD^2 = 4g \cdot J \ [\mathrm{Nm}^2] \tag{1.10}$$

where

GD^2 [Nm^2] is the flywheel moment

J [Nm $\cdot$ s^2] is the moment of inertia

G [N] is the weight

D [m] is the diameter of gyration. and

g is the acceleration of gravity: $g = 9.80665 \cong 9.81$ [m s^{-2}]

Note: when the weight G is given in kg_f, the relationship (equation (1.9)) would be:

$$J = \frac{GD^2}{4} \ [\text{kg}_f \cdot \text{m}^2] \tag{1.11}$$

Using the flywheel moment formula (equation (1.9)) and the speed in n [rpm], the equation of motion (equation (1.4)) becomes:

$$T_m = T_r + \frac{GD^2}{4g}\frac{d}{dt}\left(\frac{2\pi \cdot n}{60}\right) = T_r + \frac{GD^2}{375}\frac{dn}{dt} \ [\text{Nm}] \tag{1.12}$$

Using the English (Imperial) system of measurement
The moment of inertia is:

$$J = M \cdot \rho^2 = \frac{W}{g} \cdot R^2 \ [\text{lb} \cdot \text{ft}^2 \equiv \text{lb} \cdot \text{ft} \cdot \text{s}^2] \tag{1.13}$$

And the flywheel moment is:

$$WR^2 = g \cdot J \ [\text{lb} \cdot \text{ft}^2] \tag{1.14}$$

where

W [lb] is the weight

R [ft] is the radius of gyration, and

g [ft s^{-2}] is the acceleration of gravity.

Table 1.1. Several conversion factors.

Quantity	SI units	Converting to English units
Mass	1 [kg$_m$]	= 2.204 62 [lb$_m$]
Length	1 [m]	= 3.2808 [ft]
Acc. of gravity	~9.81 [m s^{-2}]	= 9.80665 $\cdot$ 3.2808 = 32.174 [ft s^{-2}]
Force	1 [N]	= 2.20462/9.806 65 = 1/4.4482 = 0.224 81 [lb]
Torque	1 [Nm]	= 0.224 81 $\cdot$ 3.2808 = 0.737 56 [lb $\cdot$ ft]
Moment of inertia	1 [kg$_m$ $\cdot$ m^2]	= 2.204 62 $\cdot$ 3.2808^2 = 23.73 [lb $\cdot$ ft^2]
Flywheel moment	1 [kg$_f$ $\cdot$ m^2]	= (2.204 62 $\cdot$ 3.2808^2)/4 = 5.9325 [lb $\cdot$ ft^2]

1.3 Typical speed–torque curves

The electric motor and the mechanism exhibit different speed–torque curves. When those two are connected mechanically, a stable operating point must be reached to attain steady state operation.

1.3.1 Speed–torque curves of electric motors

When analyzing the performance of an electric motor that was selected to drive a mechanism, one of the main problems is to determine whether the speed–torque curve of the motor is adequate for the requirements imposed by the speed–torque curve of the driven unit. The speed–torque curve is defined as the relationship between the speed versus torque:

$$\omega = f(T) \tag{1.15}$$

where
ω [s^{-1}] is the radial speed of the motor shaft, and
T [Nm] is the shaft-torque

The speed–torque curves of electric motors $\omega_m = f(T_m)$ can be classified into three main groups (figure 1.2):

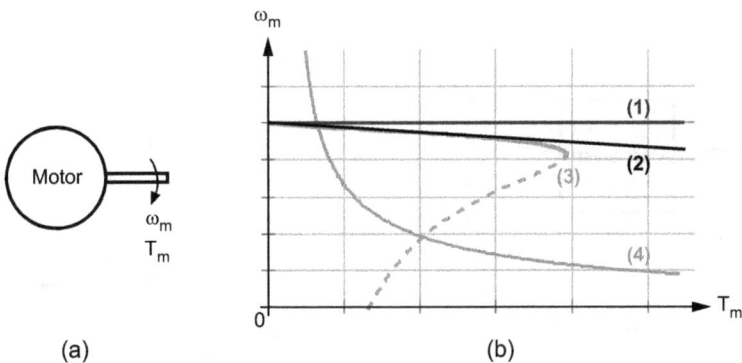

Figure 1.2. Typical speed–torque curves of several electric motors: (a) motor and (b) speed vs torque curves of four different motors.

1st group: *Flat curve* where the speed does not change with a change in torque. Synchronous motors exhibit such a curve (**curve** 1).

2nd group: *Hard curve* where the speed changes slightly with a change in torque. Two examples exhibit such a curve: a shunt-connected DC motor also a brushless DC motor (**curve 2**), and an induction motor within its actual operating region (**bold line of curve** 3).

3rd group: *Soft curve* where the speed changes significantly with a change in torque. Series-connected DC motors exhibit such a curve (**curve** 4).

1.3.2 Speed–torque curves of mechanisms

The speed versus torque curve of typical mechanisms can be expressed by an empirical equation [5]:

$$
\left.
\begin{aligned}
T_r &= T_0 + (T_n - T_0)\left(\frac{\omega_r}{\omega_n}\right)^i \quad [\text{Nm}] \\
&\text{or} \\
\omega_r &= \omega_n \cdot \sqrt[i]{\frac{T_r - T_0}{T_n - T_0}} \quad [1\ \text{s}^{-1}]
\end{aligned}
\right\}
\tag{1.16}
$$

where
  T_r [Nm] is the resistive torque required by the mechanism
  T_0 [Nm] is a resistive friction torque
  T_n [Nm] is the rated resistive torque
  ω_r [1/sec] is the speed required by the mechanism
  ω_n [1/sec] is the rated speed, and
  i is an integer. In practice, the values are: $i = -1,\ 0,\ 1,$ and 2.

The mechanism speed versus its torque $\omega_r = f(T_r)$ is shown in figure 1.3.

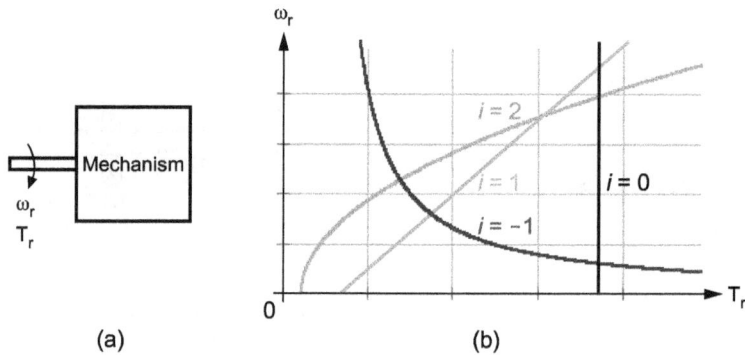

Figure 1.3. Typical speed–torque curves of mechanisms. (a) Mechanism (working machine) and (b) speed versus torque curves.

  The curves (figure 1.3) show the behavior of four typical mechanisms where the integer i varies (equation (1.16)):
  $i = -1$: A hyperbolic characteristic of machine tools such as lathes, boring, and milling machines.
  $i = 0$: A constant torque characteristic where the torque is independent of the speed, such as in crane machines during hoisting, elevators, and winches.
  $i = 1$: A linear characteristic such as in viscous friction.
  $i = 2$: A parabolic characteristic of mechanisms such as boat propellers, fans, and centrifugal pumps.

1.3.3 Stable operating point of electromechanical drives

Consider an electric motor coupled to a mechanism (figure 1.4(a)). At steady state mode of operation, the joint (operating) point of the two curves, (T_1, ω_1) in figure 1.4 (b), corresponds to a condition of balance between the driving torque of the motor and the resistive torque required by the mechanism.

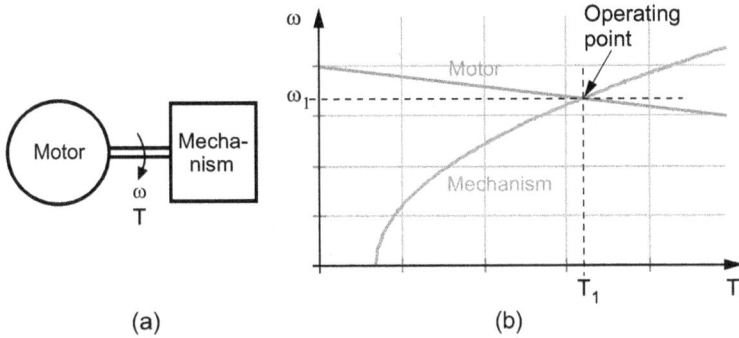

Figure 1.4. Operating point of electromechanical drives. (a) Coupled motor and mechanism, (b) speed–torque curves.

At the operating point (figure 1.4(b)), a small perturbation in speed, $\Delta\omega$, would correspond to small displacements in torques, ΔT_m in motor torque and ΔT_r in resistive torque, from the equilibrium. Substituting those changes into the equation of motion (equation (1.4)):

$$T_m + \Delta T_m = T_r + \Delta T_r + J\frac{d\omega}{dt} + J\frac{d}{dt}\Delta\omega \tag{1.17}$$

Equation (1.17) suggests:

$$\Delta T_m = \Delta T_r + J\frac{d}{dt}\Delta\omega \tag{1.18}$$

For infinitesimal changes, one can assume:

$$\frac{\Delta T_m}{\Delta\omega} \cong \frac{dT_m}{d\omega} \text{ and } \frac{\Delta T_r}{\Delta\omega} \cong \frac{dT_r}{d\omega} \tag{1.19}$$

From equations (1.18) and (1.19):

$$\left.\begin{array}{c} \dfrac{dT_m}{d\omega}\Delta\omega = \dfrac{dT_r}{d\omega}\Delta\omega + J\dfrac{d}{dt}\Delta\omega \\ \text{and} \\ J\dfrac{d}{dt}\Delta\omega + \left(\dfrac{dT_r}{d\omega} - \dfrac{dT_m}{d\omega}\right)\Delta\omega = 0 \end{array}\right\} \tag{1.20}$$

The solution for the above differential equation (equation (1.20)) is:

$$\Delta\omega = A \cdot e^{-\frac{B}{J}t} \ [\text{s}^{-1}]$$
$$\left. \text{where } B = \frac{dT_r}{d\omega} - \frac{dT_m}{d\omega} \ [\text{Nm} \cdot \text{s}] \right\} \qquad (1.21)$$
$$\text{and } A \text{ is a coefficient}$$

Equation (1.21) implies that the condition for exponential decay in $\Delta\omega$, which means a decay in the speed perturbation, must be: $B > 0$. That is:

$$\left. \begin{array}{c} B = \dfrac{dT_r}{d\omega} - \dfrac{dT_m}{d\omega} > 0 \\[1.5em] \text{or} \\[1em] \dfrac{dT_r}{d\omega} > \dfrac{dT_m}{d\omega} \end{array} \right\} \qquad (1.22)$$

Conclusion: For a stable operation, the above condition (equation (1.22)) suggests that at the operating point, the tangent to the resistive torque curve must be larger than the tangent to the driving torque curve.

Example 1.2 An electric motor is coupled to a mechanism (figure 1.4(a)). The motor develops a driving torque given by: $T_m = A - a \cdot \omega_m$. The mechanism develops a resistive torque given by: $T_r = B + b \cdot \omega_r^2$. Both curves are presented in figure 1.4(b) where the operating point is $(T_1, \ \omega_1)$.

Assume that all four coefficients, A, a, B, and b, are positive value constants, and answer:

(1) What is the speed at the operating point ?

(2) Is that a stable operating point ?

Solution

(1) At the operating point $T_1 = T_m = T_r$ and $\omega_1 = \omega_m = \omega_r$. Therefore:

$$T_1 = A - a \cdot \omega_1 = B + b \cdot \omega_1^2 \implies b \cdot \omega_1^2 + a \cdot \omega_1 + (B - A) = 0$$

and the solution for the quadratic equation is:

$$\omega_1 = \frac{-a \pm \sqrt{a^2 - 4 \cdot b(B - A)}}{2b}$$

The condition for a rational value of speed requires that positive expression under the radical, that is: $a^2 - 4 \cdot b(B - A) > 0$, which suggests that $A > B$. In addition, the condition for a positive rational value of speed, $\omega_1 > 0$, requires that the sign of the radical must be positive. Those two conditions imply that the actual speed at the operating point should be:

$$\omega_1 = \frac{-a + \sqrt{a^2 + 4 \cdot b(A - B)}}{2b} > 0$$

where $A > B$

(2) At the operation point (T_1, ω_1), the tangent to the curve of the resistive torque is:

$$\frac{\mathrm{d}T_r}{\mathrm{d}\omega} = 2b \cdot \omega_1 = -a + \sqrt{a^2 + 4 \cdot b(A - B)} > 0$$

and the tangent to the curve of the motor torque is:

$$\frac{\mathrm{d}T_m}{\mathrm{d}\omega} = -a < 0$$

Consequently: $\frac{\mathrm{d}T_r}{\mathrm{d}\omega} > \frac{\mathrm{d}T_m}{\mathrm{d}\omega}$, which indicates on a stable operating point (equation (1.22)).

1.4 Problems

1. Two identical weights, $G = 150$ [N], are connected with a cable around a pulley (see figure P1.1)

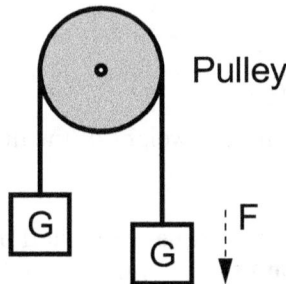

Figure P1.1. A pulley and two weights.

Neglect the system friction and the weights of the pulley and the cable, and calculate the required force F that would accelerate both weights at $a = 1$ [m s^{-2}].

2. At standstill, a weight G_1 [N] is added on top of one of the weights (figure P1.1). It causes the system to accelerate at $a_1 = 1.5$ [m s^{-2}]. Calculate the added weight ($G_1 = ?$).

3. The diameter of the pulley (figure P1.1) is $D = 0.75$ [m]. It is made of solid material and weighs 20 [kg]. As in problem 1, the two weights accelerate at $a = 1$ [m s^{-1}]. Calculate the radial acceleration of the pulley.

4. A drive system utilizes a drum and a pully to elevate a load (figure P1.4). The drum is driven by a prime mover (not shown in the figure) in a clockwise direction at a speed of n [rpm]. The pulley rotates in the counter clockwise direction with the help of a cable and elevates the load $G = 3000$ [kg] at a velocity v [m s^{-1}]. The diameter of both, the drum and the pulley, is the same: $D = 0.75$ [m].

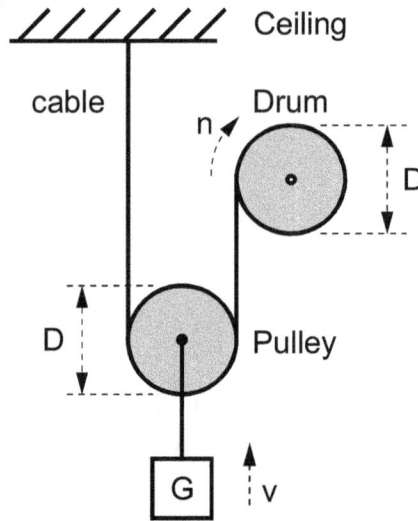

Figure P1.4. A drive system having a drum and a pulley.

Neglect the friction and the weight of the drum, the pulley and the cable, and answer:

(a) What is the required torque at the shaft of the drum to elevate the load at an acceleration of $a = 1$ [m s^{-2}]?
(b) What is the drum speed n when the load moves up at a constant velocity of $v = 0.5$ [m s^{-1}]?

References

[1] Mohan N 2001 *Electric Drives: An Integrative Approach* (MNPERE Publisher)
[2] Dubey G K 2001 *Fundamentals of Electrical Drives* (Alpha Science International Ltd)
[3] Nasar S A and Unnewehr L E 1983 *Electromechanical and Electric Machines* (New York: Wiley)
[4] Meyers R A (ed) 2002 *Encyclopedia of Physical Science and Technology* **vol 5** 3rd edn (New York: Academic)
[5] Chilikin M 1976 *Electric Drive* (MIR Publishers)
[6] Zwillinger D 2012 *Standard Mathematical Tables and Formulae* 32nd edn (Boca Raton, FL: CRC Press)

Part I

The mechanical part

The mechanical part, also called the working machine or simply mechanism, is a general term given to such equipment as machine tools, hoists, winches, cranes, conveyors, elevators, pumps, tool machines, fans, and electrical traction systems. To adapt an electric motor to the working machine specifications, all mechanical motions must be referred to the motor shaft. This part of the book addresses methods of referring steady state and dynamic motions to one shaft and provides an analysis of the transient conditions of the system dynamics.

Fundamentals of Electromechanical Drives

Zivan Zabar

Chapter 2

Referring mechanical motions

The electric motor usually drives the mechanism through a transmission apparatus where individual parts might operate at different angular or linear speeds. To join a motor and a mechanism, it is expedient to refer all torques and moments of inertia of those individual parts to the motor shaft [1–5]. This chapter addresses the translation of motion from one element to the other.

2.1 Referring torque—rotational to rotational motion

Consider an electric motor driving a working machine (mechanism) through a mechanical power transmission system (figure 2.1).

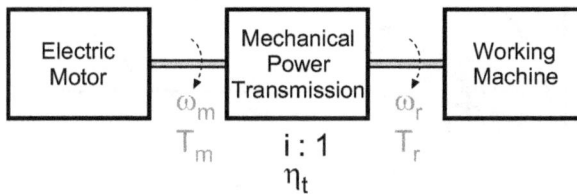

Figure 2.1. Joined motor and mechanism (working machine).

The mechanical power transmission (figure 2.1) has a speed ratio of:

$$i = \frac{\omega_m}{\omega_r} \tag{2.1}$$

where
ω_m is the radial speed of the motor shaft
ω_r is the radial speed at the load
i is the transmission ratio.

doi:10.1088/978-0-7503-6104-0ch2

The motor supplies the power (equation (1.7)) required by the mechanism and the power losses in the transmission system. The power balance is:

$$T_m \cdot \omega_m \cdot \eta_t = T_r \cdot \omega_r \; [\text{W}] \tag{2.2}$$

And the referred torque at the motor shaft would be:

$$T_m = T_r \frac{\omega_r}{\omega_m} \frac{1}{\eta_t} = \frac{T_r}{i \cdot \eta_t} \; [\text{Nm}] \tag{2.3}$$

where

T_m is the load-torque referred to the motor shaft
T_r is the resistive torque required by the mechanism (load)
η_t is the efficiency of the transmission system.

Equation (2.3) suggests that the higher the transmission ratio, the smaller the referred (required) torque at the motor shaft. Also, that referred torque is a function of the efficiency of the transmission system. The efficiency itself depends on the speed and on the delivered torque.

At a constant speed, the efficiency η improves as the torque T increases (figure 2.2), and it peaks around the rated torque $T = T_n$. At a given torque, for instance at $T = 0.6 \cdot T_n$, the efficiency varies with the speed.

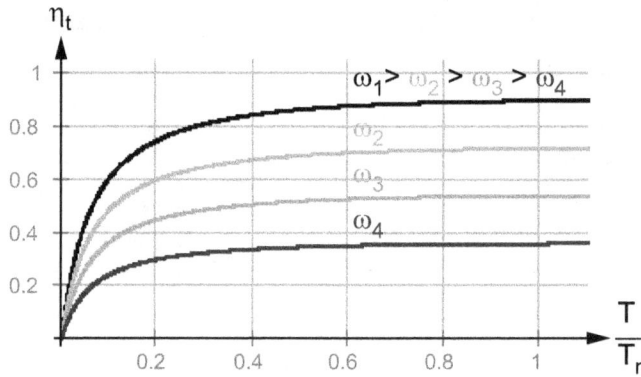

Figure 2.2. Efficiency as a function of torque.

In a braking mode of operation (see section 4.4), an active load can provide the driving torque while the motor acts as a brake. Two examples of active loads are: a gravitational force developed in hoists or elevators, and a car operating on gradients. In those cases, the load-torque covers the losses of the transmission system. The power balance is:

$$T_m \cdot \omega_m = T_r \cdot \omega_r \cdot \eta_t \; [\text{W}] \tag{2.4}$$

and the referred torque at the motor shaft is

$$T_m = T_r \frac{\omega_r}{\omega_m} \eta_t = \frac{T_r}{i} \eta_t \ [\text{Nm}] \tag{2.5}$$

The drive might include several stages of transmission between the electric motor and the mechanism (figure 2.3).

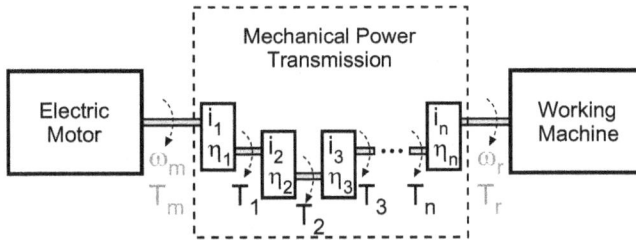

Figure 2.3. Transmission having several stages.

Each transmission stage (figure 2.3) has its own efficiency and speed ratio. At a steady-state operation, using equation (2.3), the referred torque at the motor shaft would be:

$$
\left.
\begin{array}{l}
\left.
\begin{array}{l}
T_m = \dfrac{T_1}{i_1 \cdot \eta_1} \\[2mm]
\text{and } T_1 = \dfrac{T_2}{i_2 \cdot \eta_2}
\end{array}
\right\}
\left.
\begin{array}{l}
T_m = \dfrac{T_2}{i_1 i_2 \cdot \eta_1 \eta_2} \\[2mm]
\text{and } T_2 = \dfrac{T_3}{i_3 \cdot \eta_3}
\end{array}
\right\}
T_m = \dfrac{T_3}{i_1 i_2 i_3 \cdot \eta_1 \eta_2 \eta_3} \\[4mm]
\text{Finally:} \\[2mm]
T_m = \dfrac{T_r}{i_1 i_2 i_3 \ \dots \ i_n \cdot \eta_1 \eta_2 \eta_3 \cdot \ \dots \eta_n} = \dfrac{T_r}{\displaystyle\prod_{i=1}^{n} (i_i \cdot \eta_i)} \ [\text{Nm}]
\end{array}
\right\}
\tag{2.6}
$$

Example 2.1 A hoist machine uses a winding drum that lifts and lowers a load G [N]. The drum is driven by an electric motor via three mechanical transmission stages (figure E2.1). The diameter of the drum is D, and its efficiency is η_D. The load itself moves up or down at constant velocity v.

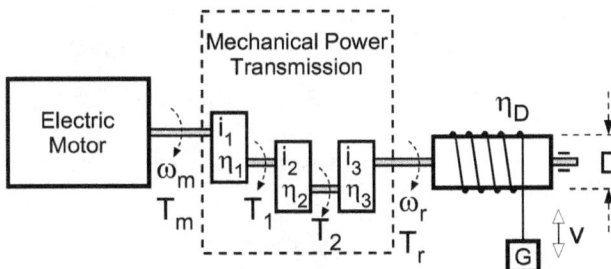

Figure E2.1. An electric hoist lift.

2-3

At a steady state operation, obtain:
  (a) the torque required at the motor shaft when the load is elevated.
  (b) the torque required at the motor shaft when the load is moving down.
  (c) the descending torque in (b) in terms of the elevating torque in (a).
  (d) the condition at which the motor must develop a downward torque to move
      the load down.

Solution
  (a) When the load is elevated, the resistive torque T_r is:

$$T_r \cdot \omega_r \cdot \eta_D = G \cdot v = G \cdot \omega_r \frac{D}{2} \ [\text{W}] \implies T_r = G\frac{D}{2} \cdot \frac{1}{\eta_D} \ [\text{Nm}]$$

The referred elevating torque at the motor shaft is (equation (2.3)):

$$\uparrow T_m = G\frac{D}{2} \frac{1}{i_1 i_2 i_3 \cdot \eta_1 \eta_2 \eta_3 \eta_D} \ [\text{Nm}]$$

  (b) When the load moves down, the resistive torque T_r is:

$$T_r \cdot \omega_r = G \cdot v \cdot \eta_D = G \cdot \omega_r \frac{D}{2} \cdot \eta_D \ [\text{W}] \implies T_r = G\frac{D}{2} \cdot \eta_D \ [\text{Nm}]$$

The referred descending torque at the motor shaft is (equation (2.5)):

$$\downarrow T_m = G\frac{D}{2} \cdot \frac{1}{i_1 i_2 i_3} \cdot \eta_1 \eta_2 \eta_3 \eta_D \ [\text{Nm}]$$

  (c) In terms of the elevating torque, the descending torque would be:

$$\downarrow T_m = \uparrow T_m \cdot (\eta_1 \eta_2 \eta_3 \eta_D)^2 \ [\text{Nm}]$$

  (d) To maintained constant velocity during a descending motion, a mechanical
      braker might be used. In practice, the motor replaces that mechanical
      braker by developing an elevating torque (a positive torque) equals the
      referred downward load-torque.
      Define:

$$i_T = i_1 i_2 i_3 \quad \text{and} \quad \eta_T = \eta_1 \eta_2 \eta_3$$

The torque losses ΔT in the transmission system can be calculated during
lifting operation:

$$\Delta T = T_r \frac{1}{i_T \cdot \eta_T} - T_r \frac{1}{i_T} \text{ [Nm]}$$

T_r was calculated in (a), and $T_r(1/i_T)$ is the ideal (no losses) referred load-torque at the motor shaft.

When *the load moves down*, the referred load-torque can be expressed in terms of the torque losses ΔT:

$$\downarrow T_m = T_r \frac{1}{i_T} - \Delta T = T_r \frac{1}{i_T} - \left(T_r \frac{1}{i_T \cdot \eta_T} - T_r \frac{1}{i_T} \right) = T_r \frac{1}{i_T} \left(2 - \frac{1}{\eta_T} \right)$$

The above equation suggests that the motor develops positive (elevating) torque as long as the efficiency is above 50% ($\eta > 0.5$). It means that the motor provides a braking torque.

When the efficiency equals or smaller than 50% ($\eta \leqslant 0.5$), the motor must develop a negative (descending) torque to move the load down.

2.2 Referring moment of inertia—rotational to rotational motion

An electric motor drives a working machine through a gearbox (figure 2.4). The moment of inertia can be delt with on the basis that the total amount of kinetic energy stored in the moving parts remains unchanged and can be referred to the motor shaft as a single moment of inertia. The total kinetic energy (equation (1.8)) referred to the motor shaft would be:

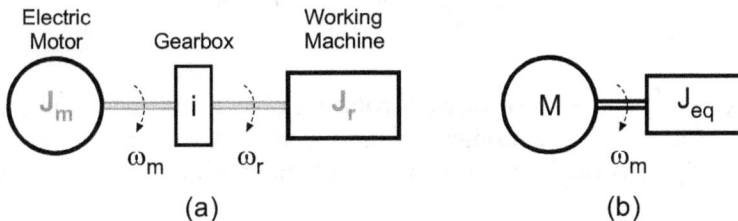

Figure 2.4. Referring moment of inertia. (a) Motor, gearbox, and working machine. (b) Motor and referred moment of inertia.

$$E_{\text{Total}} = \frac{1}{2} J_{eq} \omega_m^2 = \frac{1}{2} J_m \omega_m^2 + \frac{1}{2} J_r \omega_r^2 \text{ [joule]} \tag{2.7}$$

where

J_{eq} [Nm· s^2] is the equivalent moment of inertia at the motor shaft
J_m [Nm· s^2] is the moment of inertia of the motor itself
J_r [Nm· s^2] is the moment of inertia of the working machine

ω_m [s^{-1}] is the speed of the motor, and
ω_r [s^{-1}] is the speed of the working machine.

From equation (2.7), the equivalent moment of inertia is:

$$J_{eq} = J_m + J_r \frac{1}{i^2} \ [\text{Nm} \cdot s^2] \tag{2.8}$$

where $i = \omega_m/\omega_r$ is the gearbox ratio

Consider an electromechanical drive system having multiple rotating objects.

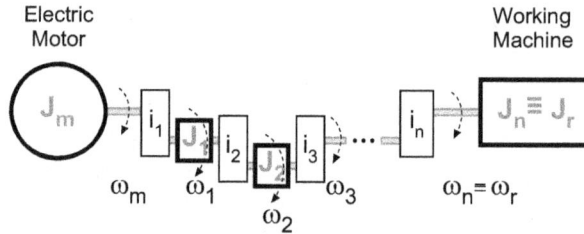

Figure 2.5. Several stages of moment of inertia at variable speeds.

Each moment of inertia of the drive (figure 2.5) rotates at a different speed. Using equation (2.8), the total kinetic energy referred to the motor shaft would be:

$$\left.\begin{aligned}
E_{\text{Total}} &= \frac{1}{2}J_{eq}\omega_m^2 = \frac{1}{2}J_m\omega_m^2 + \frac{1}{2}J_1\omega_1^2 + \frac{1}{2}J_2\omega_2^2 + \cdots + \frac{1}{2}J_n\omega_n^2 \ [\text{joule}] \\
&\text{or} \\
E_{\text{Total}} &= \frac{1}{2}J_m \cdot \omega_m^2 + \frac{1}{2}\sum_{i=1}^{n}J_i \cdot \omega_i^2 \ [\text{joule}]
\end{aligned}\right\} \tag{2.9}$$

where

$\omega_{i=1,\,2,\,3\ldots}$ [s^{-1}] is the speed of each rotating object (not including the motor's)
$J_{i=1,\,2,\,3\ldots}$ [Nm· s^2] is the moment of inertia of each rotating object, and
$J_n \equiv J_r$ [Nm· s^2] is the moment if inertia of the working machine itself.

From equation (2.9), the equivalent moment of inertia J_{eq} referred to the motor shaft would be:

$$\left.\begin{aligned}
J_{eq} &= J_m + J_1\left(\frac{\omega_1}{\omega_m}\right)^2 + J_2\left(\frac{\omega_2}{\omega_m}\right)^2 + \cdots + J_n\left(\frac{\omega_n}{\omega_m}\right)^2 \\
&= J_m + J_1\frac{1}{i_1^2} + J_2\frac{1}{i_1^2 \cdot i_2^2} + \cdots + J_n\frac{1}{i_1^2 \cdot i_2^2 \cdots i_n^2} \ [\text{Nm} \cdot s^2] \\
&\text{or} \\
J_{eq} &= J_m + \sum_{i=1}^{n}J_i\frac{1}{\left(\frac{\omega_m}{\omega_i}\right)^2} \ [\text{Nm} \cdot s^2]
\end{aligned}\right\} \tag{2.10}$$

where $i_1 = \omega_m/\omega_1$, $i_2 = \omega_1/\omega_2$, $i_3 = \omega_2/\omega_3$, ...

The equivalent flywheel moment (equation (1.10)) referred to the motor shaft would be:

$$
\left.
\begin{aligned}
(GD^2)_{eq} &= (GD^2)_m + (GD^2)_1\left(\frac{n_1}{n_m}\right)^2 + (GD^2)_2\left(\frac{n_2}{n_m}\right)^2 + \cdots + (GD^2)_n\left(\frac{n_n}{n_m}\right)^2 \\
&= (GD^2)_m + (GD^2)_1\frac{1}{i_1^2} + (GD^2)_2\frac{1}{i_1^2 \cdot i_2^2} + \cdots + (GD^2)_n\frac{1}{i_1^2 \cdot i_2^2 \cdots i_n^2} \ [\text{Nm}^2] \\
&\hspace{5cm} or \\
&= (GD^2)_m + \sum_{i=1}^{n}(GD^2)_i\frac{1}{\left(\dfrac{n_m}{n_i}\right)^2} \ [\text{Nm}^2]
\end{aligned}
\right\}
\quad (2.11)
$$

where $i_1 = n_m/n_1$, $i_2 = n_1/n_2$, $i_3 = n_2/n_3$, ...

2.3 Referring force—linear to rotational motion

Consider an electromechanical drive utilizing a winding drum that lifts and lowers a load (figure 2.6). The linear motion of the load can be referred to a rotating shaft on the basis of the power balance in the system.

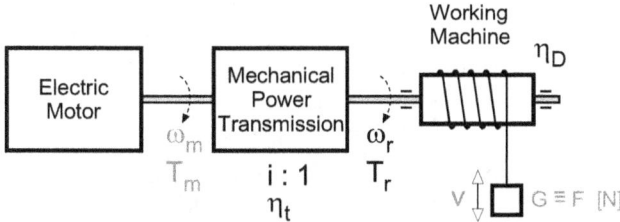

Figure 2.6. An electromechanical drive utilizing a winding drum (referring force).

When the load G [N] is elevated at a velocity of $v[\text{m s}^{-1}]$, the power balance is:

$$T_r \cdot \omega_r \cdot \eta_D = F \cdot v \ [\text{W}] \quad (2.12)$$

where
   v is the velocity of the load,
   F is the gravitational force, which is the weight of the load G in [N], and
   η_D is the efficiency of the winding drum.

The referred torque at the motor shaft is (equations (2.3) and (2.12)):

$$T_m = T_r\frac{1}{i \cdot \eta_t} = \frac{F \cdot v}{\omega_r \cdot \eta_D} \cdot \frac{1}{i \cdot \eta_t} = \frac{F \cdot v}{\omega_m \cdot \eta} \ [\text{Nm}] \quad (2.13)$$

where $\omega_m = i \cdot \omega_r$ (equation (2.1)), and $\eta = \eta_t \cdot \eta_D$.

2.4 Referring force—linear to linear motion

Consider a mechanical drive utilizing a rack and pinion gear that lifts and lowers a load.

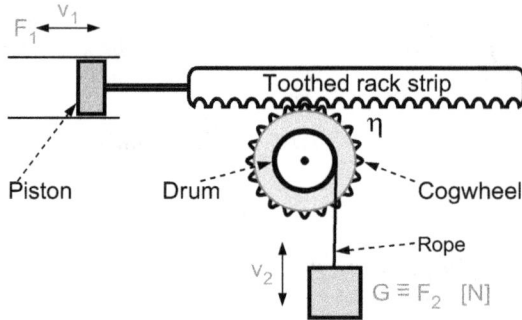

Figure 2.7. A mechanical drive utilizing a rack and pinion gear.

The reciprocating piston drives the toothed rack strip at a force of F_1 [N] and at a velocity of v_1 [m s^{-1}] (figure 2.7). The rack rotates the cogwheel at an efficiency η. The drum is attached to the cogwheel shaft and lifts or lowers the load $G \equiv F_2$ [N] at a velocity of v_2 [m s^{-1}]. The power balance is:

$$F_1 \cdot v_1 \cdot \eta = F_2 \cdot v_2 \text{ [W]} \tag{2.14}$$

The referred force F_1 at the reciprocating piston is:

$$F_1 = \frac{F_2 \cdot v_2}{v_1 \cdot \eta} \text{ [N]} \tag{2.15}$$

2.5 Referring mass—linear to rotational motion

Consider an electromechanical drive utilizing a winding drum that lifts and lowers a load.

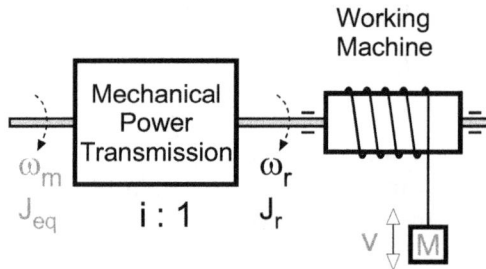

Figure 2.8. An electromechanical drive utilizing a winding drum (referring mass).

The mass (figure 2.8) can be referred to (translated to) a rotating shaft on the basis of its kinetic energy (equations (1.3) and (1.8)). Namely, when the mass M moves at a linear velocity v, its equivalent kinetic energy referred to the shaft of the working machine itself would be:

$$\frac{1}{2}J_r \cdot \omega_r^2 = \frac{1}{2}M \cdot v^2 \quad \text{[joule]} \tag{2.16}$$

and the equivalent moment of inertia J_r at the shaft of the working machine itself is:

$$J_r = M\left(\frac{v}{\omega_r}\right)^2 \quad [\text{Nm} \cdot \text{s}^2] \tag{2.17}$$

Finally, the equivalent moment of inertia J_{eq} (figure 2.8), referred to the moving mass M, would be:

$$J_{eq} = J_r \frac{1}{i^2} = M\left(\frac{v}{\omega_m}\right)^2 \quad [\text{Nm} \cdot \text{s}^2] \tag{2.18}$$

Substituting the flywheel moment expression (equation (1.10)) for the moment of inertia J_{eq} in equation (2.18):

$$\frac{(GD^2)_{eq}}{4g} = \frac{G}{g}\left(\frac{v}{\dfrac{2\pi \cdot n_m}{60}}\right) \tag{2.19}$$

where
G is the load: $G = M \cdot g$ [N],
n_m [rpm] is the speed of the motor shaft, and
g is the acceleration of gravity: $g \cong 9.81$ [m s^{-1}].

Finally, the equivalent flywheel moment at the motor shaft would be:

$$(GD^2)_{eq} = 365 \cdot G\left(\frac{v}{n_m}\right)^2 \quad [\text{Nm}^2] \tag{2.20}$$

2.6 Referring mass—linear to linear motion

Consider an electromechanical drive utilizing two winding drums that lift and lower their loads at different speeds (figure 2.9).

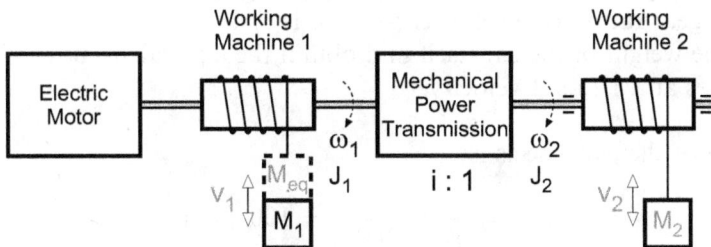

Figure 2.9. An electromechanical drive utilizing two winding drums (referring mass).

As stated in section 2.5 above, when the mass M_2 moves at a linear velocity v_2, its equivalent kinetic energy referred to the shaft of working machine 2 would be:

$$\frac{1}{2}J_2 \cdot \omega_2^2 = \frac{1}{2}M_2 \cdot v_2^2 \quad \text{[joule]} \tag{2.21}$$

and the equivalent moment of inertia J_2 at the shaft of working machine 2 is:

$$J_2 = M_2\left(\frac{v_2}{\omega_2}\right)^2 \quad \text{[Nm} \cdot \text{s}^2\text{]} \tag{2.22}$$

The equivalent moment of inertia J_1, referred to the moving mass M, at the shaft of working machine 1 is:

$$J_1 = J_2\frac{1}{i^2} = M_2\left(\frac{v_2}{\omega_1}\right)^2 \quad \text{[Nm} \cdot \text{s}^2\text{]} \tag{2.23}$$

An equivalent mass M_{eq} on top of M_1 (figure 2.9) and moves with it, would have a kinetic energy:

$$\frac{1}{2}M_{eq} \cdot v_1^2 = \frac{1}{2}J_1 \cdot \omega_1^2 \quad \text{[joule]} \tag{2.24}$$

Finally, the referred (equivalent) mass M_{eq} is:

$$M_{eq} = J_1\left(\frac{\omega_1}{v_1}\right)^2 = M_2\left(\frac{v_2}{v_1}\right)^2 \quad \text{[N} \cdot \text{s}^2 \text{ m}^{-1}\text{]} \tag{2.25}$$

Note: The equivalent mass (equation (2.25)) can also be derived directly on the basis that kinetic energy stored in the moving parts remains unchanged. That is:

$$\frac{1}{2}M_{eq} \cdot v_1^2 = \frac{1}{2}M_2 \cdot v_2^2 \implies M_{eq} = M_2\left(\frac{v_2}{v_1}\right)^2 \quad \text{[N} \cdot \text{s}^2 \text{ m}^{-1}\text{]}$$

Example 2.2 A toy car uses the kinetic energy stored in its rotating internal flywheel to propel itself. The flywheel has a shape of a cylindrical coin and it is made of a solid material. Its diameter is 4 cm and it weighs 50 g. The diameter of the car wheels is 2 cm. The gear ratio, flywheel to car-wheel, is 5: 1.

Neglect the weight of the car itself and obtain the equivalent mass of the toy car when it moves at a 1 m s^{-1} velocity.

Solution

The speed of the car wheels is:

$$\omega_C = \frac{v_C}{r} = \frac{1}{\dfrac{0.02}{2}} = 100 \quad \text{[s}^{-1}\text{]}$$

The speed of the flywheel is: $\omega_F = i \cdot \omega_C = 5 \cdot 100 = 500$ [s^{-1}]

The moment of inertia of the flywheel is (see example 1.1):

$$J_F = M\frac{R^2}{2} = 0.05\frac{\left(\dfrac{0.04}{2}\right)^2}{2} = 10^{-5} \ [\text{kg}_\text{m} \ \text{m}^2]$$

When the toy car moves at a velocity of $1\,\text{m}\,\text{s}^{-1}$, its equivalent mass is (equation (2.18)):

$$J = M\left(\frac{v}{\omega}\right)^2 \implies M_{eq} = J_F\left(\frac{\omega_F}{v}\right)^2 = 10^{-5}\left(\frac{500}{1}\right)^2 = 2.5 \ [\text{kg}_\text{m}]$$

Example 2.3 An electric motor moves a metal table via a mechanical power transmission, which is a gear system (figure E2.3.1). Cogwheel #1 is attached to the motor shaft (motor is not shown in the figure) and coupled with the three identical cogwheels #2. These three are coupled with the internal teeth of cogwheel #3. The outer teeth of cogwheel #3 moves the metal table at a velocity v (table E2.1).

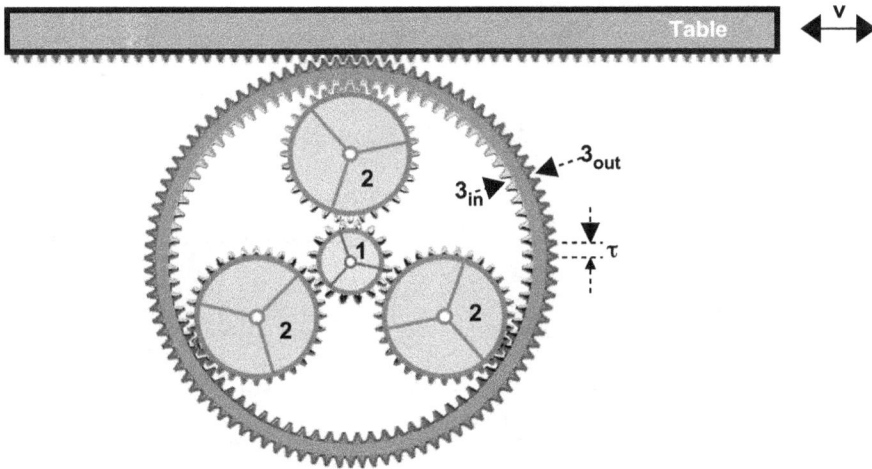

Figure E2.3.1. A mechanical gear system.

Table E2.1. Given system parameters.

	Motor and cogwheel #1	Cogwheel #2	Cogwheel #3
Flywheel moment GD^2 [Nm²]	$(GD^2)_1 = 0.46$	$(GD^2)_2 = 0.74$	$(GD^2)_3 = 16.5$
No. of teeth Z	$Z_1 = 16$	$Z_2 = 32$	$Z_3^{in} = 140$ $Z_3^{out} = 150$
Tooth pitch τ [mm]	$\tau_1 = 6.5$	$\tau_2 = 6.5$	$\tau_3^{in} = 6.5$ $\tau_3^{out} = 6.5$

The metal table weighs $G_T = 1180$ [N], and the motor speed is $n_m = 440$ rpm.

Calculate the equivalent flywheel moment at the motor shaft.

Solution

The gear ratio of two coupled cogwheels is:

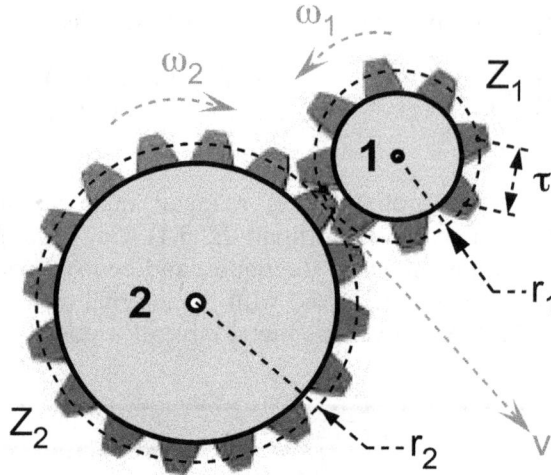

Figure E2.3.2. Two coupled cogwheels.

The tangential speed v between the two coupled cogwheels (figure E2.3.2) is:

$$v = \omega_1 \cdot r_1 = \omega_2 \cdot r_2$$

The gear ratio would be:

$$i = \frac{\omega_1}{\omega_2} = \frac{r_2}{r_1} \cdot \frac{2\pi}{2\pi} = \frac{Z_2 \cdot \tau}{Z_1 \cdot \tau} = \frac{Z_2}{Z_1}$$

where

   Z_1 and Z_2 are the number of teeth in cogwheel #1 and cogwheel #2, respectively,

   r_1 and r_2 are the radii of cogwheel #1 and cogwheel #2, respectively,

   ω_1 and ω_2 are the angular speed of cogwheel #1 and cogwheel #2, respectively, and

   τ is the tooth pitch of both cogwheels.

Based on the above, the speed of cogwheel #3 would be:

$$\left.\begin{array}{l} n_2 = \dfrac{Z_1}{Z_2}n_1 = \dfrac{n_1}{i_1} \\[2ex] n_3 = \dfrac{Z_2}{Z_3^{in}}n_2 = \dfrac{n_2}{i_2} \end{array}\right\} \quad n_3 = \frac{n_1}{i_1 \cdot i_2} = \frac{440}{\dfrac{32}{16} \cdot \dfrac{140}{32}} = 18.3 \ [\text{rpm}]$$

The linear velocity of the table would be (figure E2.3.1):

$$v = \omega_3 \cdot r_3 = \frac{2\pi \cdot n_3}{60} r_3 = Z_3^{out} \cdot \tau \frac{n_3}{60} = 150 \cdot 6.5 \cdot 10^{-3} \frac{18.3}{60} = 0.3 \; [\text{m s}^{-1}]$$

The total equivalent flywheel moment at the motor shaft would be (equations (2.11) and (2.20)):

$$(GD^2)_{eq} = (GD^2)_1 + 3 \cdot (GD^2)_2 \frac{1}{i_1^2} + (GD^2)_3 \frac{1}{i_1^2 \cdot i_2^2} + 365 \cdot G_T \left(\frac{v}{n_m}\right)^2$$

$$= 0.46 + 3 \cdot 0.74 \frac{1}{\left(\frac{32}{16}\right)^2} + 16.5 \frac{1}{\left(\frac{32}{16}\right)^2 \left(\frac{140}{32}\right)^2} + 365 \cdot 1180 \left(\frac{0.3}{440}\right)^2 = 1.431 \; [\text{Nm}^2].$$

2.7 Referring torque and mass—eccentric crank motion

Consider an eccentric motion of a mechanism, such as a crank gear (figure 2.10). Translating the speed and acceleration of the mass, the torque, and the moment of inertia requires to factor in the variations in magnitude and direction throughout each revolution of the crank.

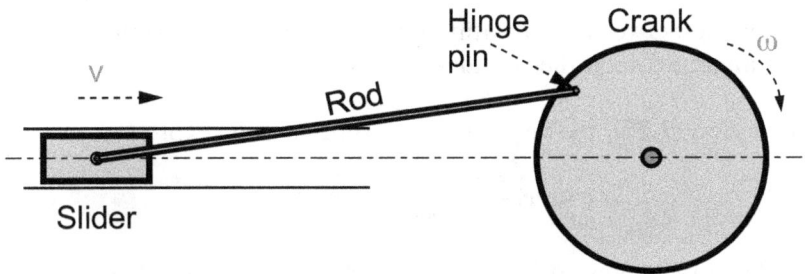

Figure 2.10. Slider crank mechanism.

The slider crank mechanism (figure 2.10) is used to convert a rotational motion into a reciprocating motion and vice versa. The mechanism is composed of three major parts: the crank, which is the rotating wheel; the slider, which slides inside a hollow cylinder; and a connecting rod, which join the crank and the slider. The total distance that is covered by the slider equals the diameter of the crank.

2.7.1 The reciprocating velocity of the slider

As the crank rotates at a speed ω, the reciprocating velocity v of the slider varies. Relating to figure 2.10, the velocity vectors are as in figure 2.11.

In figure 2.11, the velocity vectors are illustrated in red dashed lines. The length of the rod is l, the radius of the crank wheel is r, the linear velocity of the slider is v, and the distance between the shaft of the crank and the slider is x. This distance can be expressed in terms of the radius of the crank and the length of the rod:

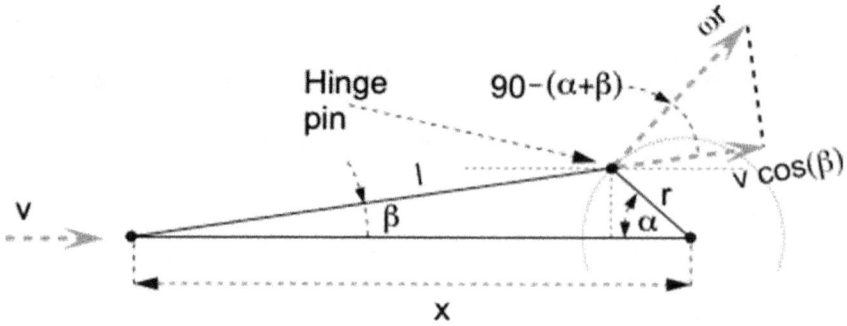

Figure 2.11. Velocity vectors for the slider crank mechanism.

$$x = \sqrt{l^2 - (r \cdot \sin \alpha)^2} + r \cdot \cos \alpha \qquad (2.26)$$

Knowing that the magnitude of the velocity vector along the rod is $v \cdot \cos(\beta)$, and that the tangential velocity of the crank is ωr, one can deduce:

$$\omega r \cdot \cos [90 - (\alpha + \beta)] = v \cdot \cos \beta \qquad (2.27)$$

where

α is the angle between the rotating hinge pin and the horizontal axis, and
β is the angle between the rod and the horizontal axis.

From equation (2.27), the velocity of the slider can be expressed as:

$$v = \omega r \frac{\sin (\alpha + \beta)}{\cos \beta} \; [\text{m s}^{-1}] \qquad (2.28)$$

The angle β can be expressed in terms of the angle α (figure 2.11):

$$\left. \begin{aligned} r \cdot \sin \alpha &= l \cdot \sin \beta \\ \beta &= \sin^{-1}\left[\frac{r}{l} \sin \alpha \right] \end{aligned} \right\} \qquad (2.29)$$

The velocity of the slider (equation (2.28)) can also be express in an alternative way. That is, using the law of cosines [6], the length of the rod is:

$$l^2 = x^2 + r^2 - 2xr \cdot \cos \alpha \qquad (2.30)$$

The distance x between the slider and the shaft of the crank varies over time as the wheel rotates. Using equation (2.30), the rate of change with respect to the time provides:

$$0 = 2x\frac{dx}{dt} - 2r \cdot \cos \alpha \frac{dx}{dt} + 2xr \cdot \sin \alpha \frac{d\alpha}{dt} \qquad (2.31)$$

From equation (2.31), the velocity of the slider can be expressed as a function of the angle α:

$$v = \frac{dx}{dt} = \frac{\omega r \cdot x \cdot \sin \alpha}{r \cdot \cos \alpha - x} \tag{2.32}$$

Substituting the distance x (equation (2.26)) in the alternate velocity formula (equation (2.32)) gives:

$$v = \omega r \cdot \sin \alpha \left[1 + \frac{\cos \alpha}{\sqrt{\left(\frac{l}{r}\right)^2 - (\sin \alpha)^2}} \right] \ [\text{m s}^{-1}] \tag{2.33}$$

Both formulas (equations (2.28) and (2.33)) would yield identical results.

2.7.2 The equivalent moment of inertia

Based on the fact that the total amount of kinetic energy stored in the moving parts remains unchanged, the translated mass of the slider can be referred to the rotating shaft of the crank:

$$\frac{1}{2} J_{eq} \cdot \omega^2 = \frac{1}{2} M \cdot v^2 \ [\text{joule}] \tag{2.34}$$

where
$M \ [\text{N} \cdot \text{s}^2 \ \text{m}^{-1}]$ is the mass of the slider,
$v \ [\text{m s}^{-1}]$ is the velocity of the slider,
$J_{eq} \ [\text{Nm} \ \text{s}^2]$ is the equivalent moment of inertia at the shaft of the crank, and
$\omega \ [\text{s}^{-1}]$ is the angular speed of the crank.

From equation (2.34), the equivalent moment of inertia can be expressed:

$$J_{eq} = M \left(\frac{v}{\omega}\right)^2 \ [\text{Nm} \cdot \text{s}^2] \tag{2.35}$$

Substituting the velocity value (equation (2.33)) in the above equation, the equivalent moment of inertia becomes:

$$J_{eq} = Mr^2 \left\{ \sin \alpha \left[1 + \frac{\cos \alpha}{\sqrt{\left(\frac{l}{r}\right)^2 - (\sin \alpha)^2}} \right] \right\}^2 \ [\text{Nm} \cdot \text{s}^2] \tag{2.36}$$

Alternatively, one can use the velocity formula given in equation (2.28). In that case, the equivalent moment of inertia becomes:

$$J_{eq} = Mr^2 \left[\frac{\sin(\alpha + \beta)}{\cos(\beta)} \right]^2 [\text{Nm} \cdot \text{s}^2] \tag{2.37}$$

Both equations (equations (2.36) and (2.37)) would yield identical results.

Substituting the flywheel moment (equation (1.10)) for the moment of inertia:

$$(GD^2)_{eq} = 4 \cdot Gr^2 \left[\frac{\sin(\alpha + \beta)}{\cos(\beta)} \right]^2 [\text{Nm}^2] \tag{2.38}$$

where

$G = M \cdot g$ [N] is the weight of the slider, and

g is the acceleration of gravity: $g = 9.80665 \cong 9.81$ [m s^{-2}].

When the crank makes one complete revolution, the characteristic of the velocity v (equation (2.33)), the distance x (equation (2.26)), and the equivalent moment of inertia J_{eq} (equation (2.36)) as a function the angle α are shown in figure 2.12.

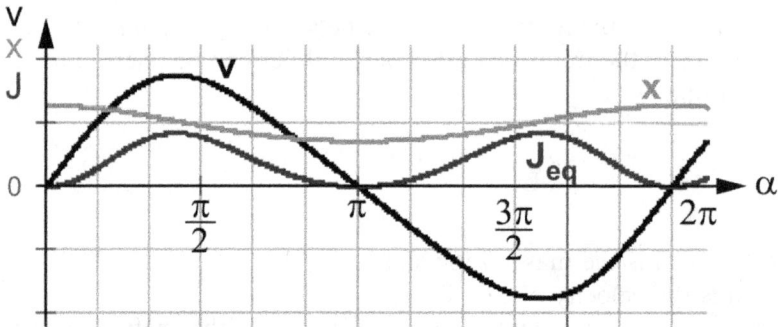

Figure 2.12. Velocity v, distance x, and moment of inertia J_{eq} as function of α.

During each cycle of the crank, the reciprocating motion of the slider is depicted by curve x (figure 2.12). The reciprocal motion of the slider is depicted by curve v, which resembles a sine wave shape. The oscillatory action of the moment of inertia, which is depicted by curve J_{eq}, peaks at maximum velocity and has a negligible value at zero velocity.

Notes:

(a) For $x \gg r$, the velocity curve would become closer to a pure sine wave.

(b) In practice, a relative heavy flywheel attached to the rotating shaft of the crank would minimize the ripple effect in the moment of inertia.

2.7.3 Energy, power, and torque

The equivalent kinetic energy stored in the rotating crank is (equation (1.8)):

$$E = \frac{1}{2}J_{eq} \cdot \omega^2 \left[\text{Nm} \cdot \text{s}^2 \cdot \frac{1}{\text{s}^2} = \text{Nm} = \text{joule} = \text{W} \cdot \text{s} \right] \tag{2.39}$$

The equivalent moment of inertia J_{eq} (equation (2.36)) varies with the angle α.
At a constant radial speed, $\omega = \text{const.}$, the related power would be:

$$P = \frac{dE}{dt} = \frac{1}{2}\omega^2 \frac{dJ_{eq}}{dt} = \frac{1}{2}\omega^2 \frac{dJ_{eq}}{d\alpha} \cdot \frac{d\alpha}{dt} = \frac{1}{2}\omega^3 \frac{dJ_{eq}}{d\alpha} \left[\frac{1}{\text{s}^3}\text{Nm} \cdot \text{s}^2 = \text{W} \right] \tag{2.40}$$

And the equivalent torque becomes:

$$T_{eq}^{\alpha} = \frac{P}{\omega} = \frac{1}{2}\omega^2 \frac{dJ_{eq}}{d\alpha} \left[\frac{1}{\text{s}^2}\text{Nm} \cdot \text{s}^2 = \text{Nm} \right] \tag{2.41}$$

At a variable radial speed, $\omega = \text{var.}$, the related power would be:

$$P = \frac{dE}{dt} = J_{eq}\omega\frac{d\omega}{dt} + \frac{1}{2}\omega^3 \frac{dJ_{eq}}{d\alpha} \quad [\text{W}] \tag{2.42}$$

And the equivalent torque becomes:

$$T_{eq}^{\alpha,\ \omega} = \frac{P}{\omega} = J_{eq}\frac{d\omega}{dt} + \frac{1}{2}\omega^2 \frac{dJ_{eq}}{d\alpha} \quad [\text{Nm}] \tag{2.43}$$

2.8 Problems

1. An electric motor M rotates at a speed n_m and drives a mechanical system via a gearbox. The shaft of the mechanical system includes a winding drum, a dry friction part, a viscous friction, and a fan (figure P2.1).

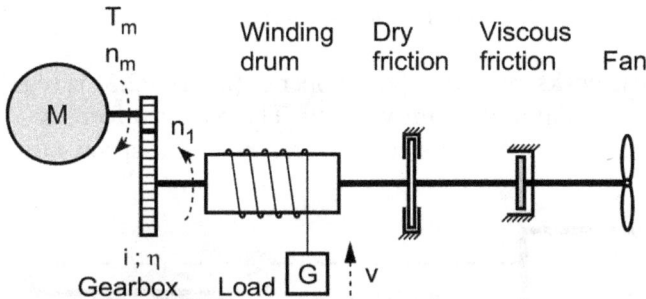

Figure P2.1. An electromechanical drive system.

The flywheel moment of the motor is $(GD^2)_m$ [Nm]. The speed ratio of the gearbox is $i = n_m/n_1$ and its efficiency is η. The diameter of the winding drum is D [m], and it has a flywheel moment of $(GD^2)_F$ [Nm²]. The drum elevates a load G [N] at a velocity of v [m s^{-1}]. The dry friction part requires a constant torque T_D [Nm]. The viscous friction machine requires a torque that relates to its shaft speed $T_V = k_v \cdot n_1$ [Nm]. The fan requires a torque that relates to the square of its speed $T_F = k_f \cdot n_1^2$ [Nm].

Neglect the weight of all components except those of the motor, the drum and the load, and calculate the total torque T_m required at the motor shaft.

2. An electric motor dirives a winding drum via a gearbox (figure P2.2). The motor speed is $n_m = 715$ [rpm]. The diameter of the winding drum is $D = 0.294$ [m] and its efficiency is $\eta_D = 0.8$. The efficiency of the gearbox is $\eta_i = 0.875$. The weight of the load is $G = 1500$ [N], and it is lifted at a constant velocity of $v = 1$ [m s^{-1}].

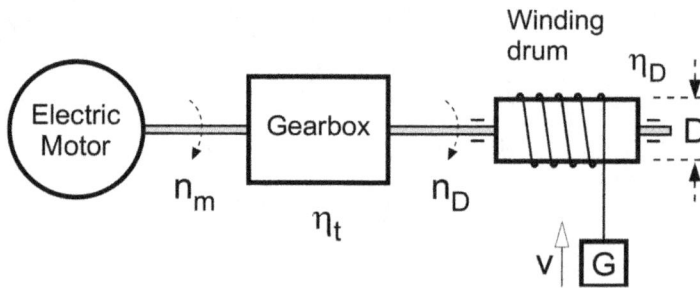

Figure P2.2. An electric motor drives a winding drum.

Calculate:
(a) The resistive torque referred to the motor shaft.
(b) The required power of the motor shaft.
(c) The equivalent flywheel moment at the motor shaft (consider only the weight of the linear moving load).

3. An electric corkscrew wine bottle opener (figure P2.3) is required to pull the cork out in 4 s at a constant velocity. The required force is $F = 300$ [N]. An electric motor drives the screw via gearbox of a speed ratio $i = 10$ (motor

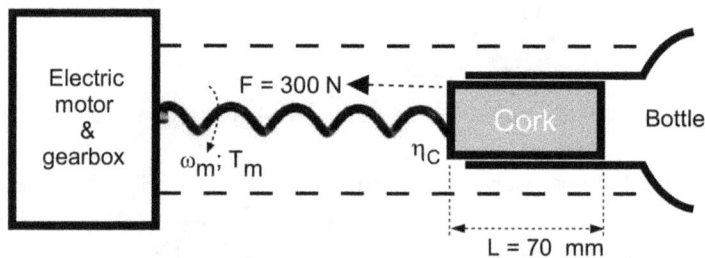

Figure P2.3. An electric wine bottle opener.

runs faster) and an efficiency of $\eta_G = 90\%$. The screw rotates at 1 revolution per second and penetrates the cork at an efficiency of $\eta_C = 10\%$.

Calculate the referred torque at the motor shaft and the required power of the motor.

4. A hoist machine consists of a driving motor, a gearbox of a speed ratio $i = 4$ (motor runs faster), a winding drum of $D_D = 0.4$ [m] diameter, a pulley of $D_P = 0.5$ [m] diameter, two cars that weigh $G_C = 100$ kg$_f$ each having a load of $G_{L1} = 500$ [kg$_f$] and $G_{L2} = 100$ [kg$_f$]. Car#1 is elevated at a linear velocity of $v = 4.6$ [m s^{-1}], while car#2 is descending. Additional data is provided in the figure (figure P2.4).

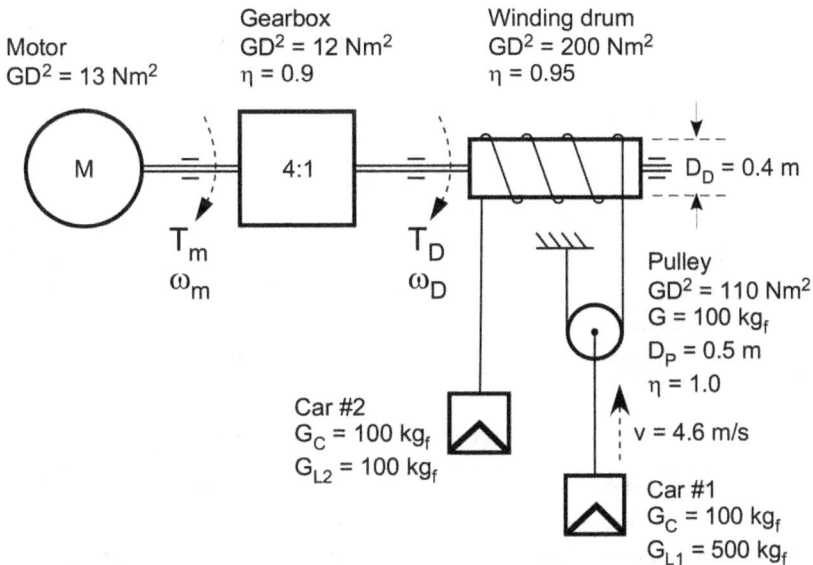

Figure P2.4. An electromechanical drive system having two cars.

Neglect the cable weight and calculate:
(a) The required power of the motor.
(b) The equivalent flywheel moment referred to the motor shaft.

5. An electric motor drives a shaping machine via a power transmission system consists of multiple cogwheel combinations. The load $G_L = 1800$ [N] is attached to a moving table that weighs $G_T = 1230$ [N]. The tooth pitch of all cogwheels is $\tau = 25.13$ [mm]. When the motor rotates at a speed of $n_m = 420$ [rpm], the table moves at a velocity of v_T [m s^{-1}]. Additional data is provided in figure P2.5.

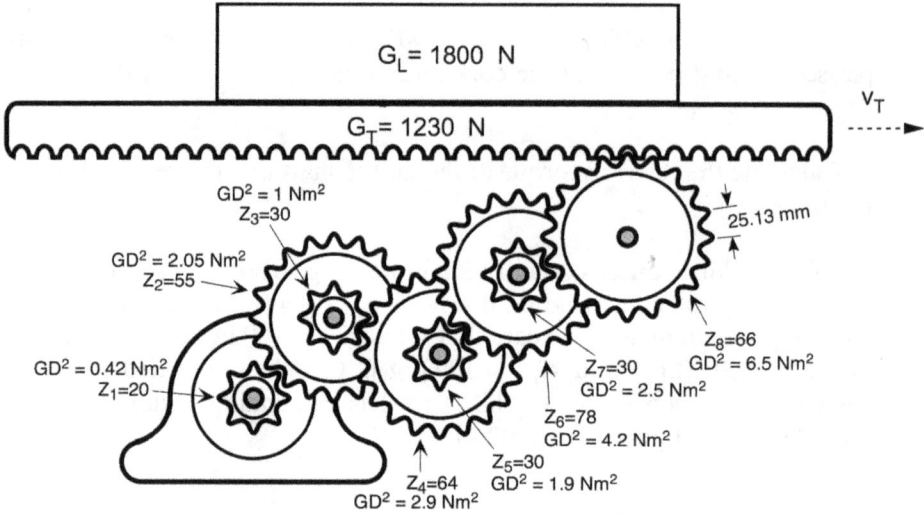

Figure P2.5. A shaping machine with multiple cogwheels.

Calculate the equivalent flywheel moment $(GD^2)_{eq}$ at the motor shaft.

6. A tool machine (shaping machine) utilizes a double gear to move a load from one side to the other (figure P2.6). The motor is coupled directly to the shaft of the gear, and it moves longitudinally (left and right) with the gear. The two cogwheels are attached directly to the motor shaft and rotate with it at the same speed.

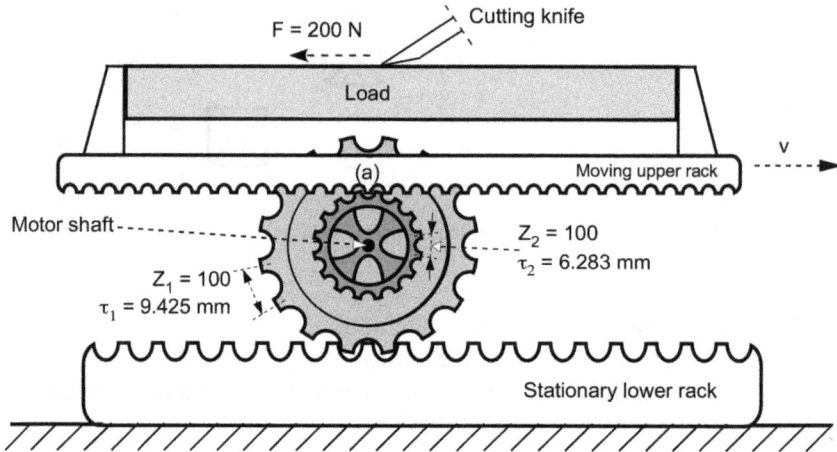

Figure P2.6. A shaping machine with double gear.

The larger cogwheel of the double gear has $Z_1 = 100$ cogs (teeth) with a cog-pitch $\tau_1 = 9.425$ [mm], and rolls on the stationary lower rack. The smaller cogwheel has the same number of cogs $Z_2 = 100$, but its cog-pitch is $\tau_2 = 6.283$ [mm], and it thrusts the upper rack with the load.

Given:
- The weight of the load including the upper rack is 1000 [N].
- The shaping tool generates a resistive force $F = 200$ [N] (assume F acts at point (a).
- The total efficiency of the gear is 90.9%.
- The weight of the smaller cogwheel is 500 [N], and its diameter of gyration is $0.7 \cdot D_2$ (where D_2 is the diameter of the small cogwheel).
- The weight of the larger cogwheel is 800 [N], and its diameter of gyration is $0.72 \cdot D_1$ (where D_1 is the diameter of the large cogwheel).
- The weight of the motor itself is 200 [N], and its flywheel moment is $(GD^2)_m = 24$ [Nm^2].
- The motor rotates at $n_m = 191$ [rpm].
   Calculate:

   (a) The required shaft-power of the motor.
   (b) The equivalent flywheel moment referred to the motor shaft.

7. An electric motor drives an industrial cart at a velocity of $v = 20$ [m s^{-1}]. The motor propels the cart via a gear system coupled to the shaft of two main wheels, out of four. The diameter of each main wheel is $D = 2$ [m]. The number of teeth Z of each cogwheel are given in the figure. Cogwheel Z_1 is attached to the motor shaft, and cogwheel Z_4 is attached to the shaft of the two driving wheels (figure P2.7).

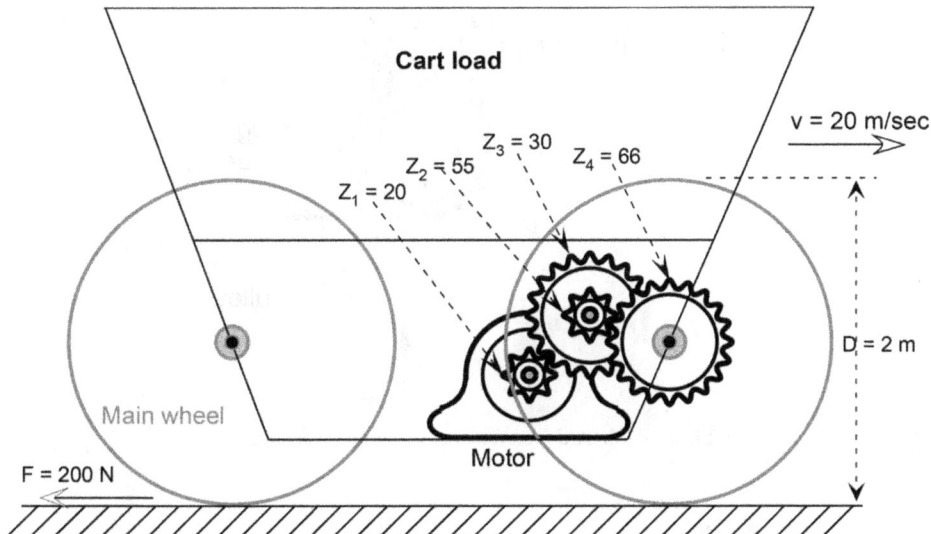

Figure P2.7. An industrial cart driven by an electric motor.

- The total weight of the cart itself is $G_C = 509.7$ [kg$_f$], and the maximum load it can carry is $G_L = 203.9$ [kg$_f$].

- The efficiency of the first gear, between Z_1 and Z_2 is $\eta_1 = 0.95$, and the efficiency of the second gear, between Z_3 and Z_4 is $\eta_2 = 0.92$.
- The friction of each main wheel and the ground is $F = 200$ [N].
- The flywheel moment of each of the four main wheels is $(GD^2)_w = 2020$ [Nm²].
- The flywheel moment of the motor and its cogwheel Z_1 is $(GD^2)_m = 52$ [Nm²].
- The flywheel moment of cogwheels Z_2 and Z_3 and shaft is $(GD^2)_{2,3} = 420$ [Nm²].
- The flywheel moment of the third (front) shaft including Z_4 is $(GD^2)_4 = 650$ [Nm²].

  Calculate:

  (a) The total flywheel moment, GD^2, transferred to the motor shaft.
  (b) The required power rating of the motor at continuous (steady-state) load.

8. An electric motor drives a warm gear to lift a load G (figure P2.8). The gear's cogwheel and the drive wheels are connected to the same shaft. The drive wheel pulls a cable that lifts a pulley with the load. The other end of the cable is anchored to the ground.

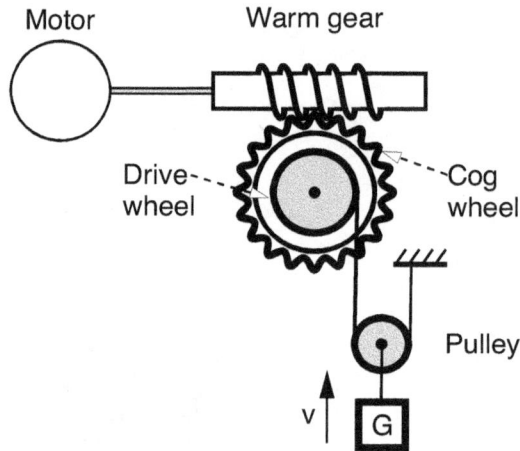

Figure P2.8. A drive system employing a warm gear.

- The flywheel moment of the motor is: $[GD^2]_m = 1$ [Nm²].
- The inertia of the warm gear and its shaft is: $[GD^2]_w = 0.2$ [Nm²].
- The efficiency of the warm gear is: $\eta = 0.9$.
- The inertia of the cogwheel is: $[GD^2]_C = 250$ [Nm²].

- Number of teeth at the rim of the cogwheel is: $Z = 32$.
- The inertia of the drive wheel is: $[GD^2]_D = 125$ [Nm^2].
- The diameter of the drive wheel is: $D_D = 1.4$ [m].
- The mass of the pulley is: $M_P = 100$ [kg$_m$].
- The diameter of the pulley is: $D_P = 0.2$ [m].
- The inertia of the pulley is given by: $J_P = M_P \cdot (D_p/2)^2$.
- The weight of the load is: $G = 200$ [kg].
- The velocity of the load is: $v = 4$ [m s^{-1}].

    Neglect the cable weight and friction at the wheels, and calculate:

    (a) the motor speed and its required power.

    (b) the equivalent flywheel moment referred to the motor shaft.

References

[1] Mohan N 2001 *Electric Drives: An Integrative Approach* (MNPERE Publisher)

[2] Dubey G K 2001 *Fundamentals of Electrical Drives* (Alpha Science International Ltd.)

[3] Nasar S A and Unnewehr L E 1983 *Electromechanical and Electric Machines* (New York: Wiley)

[4] Meyers R A (ed) 2002 *Encyclopedia of Physical Science and Technology* **vol 5** 3rd edn (New York: Academic)

[5] Chilikin M 1976 *Electric Drive* (MIR Publishers)

[6] Zwillinger D 2012 *Standard Mathematical Tables and Formulae* 32nd edn (Boca Raton, FL: CRC Press)

Chapter 3

Mechanical transient conditions

In studying the behavior of an electromechanical drive system, one of the problems is to analyze the transient conditions of the system dynamics. A transient condition occurs when a transition is made from one steady state of operation to another in which the speed and torque undergo certain changes. This chapter addresses the transient time required for those changes and provides a computational method to tackle a nonlinear motion.

3.1 Acceleration and deceleration time

An electromechanical transient process is related to the dynamic behavior of both the driving motor and the driven mechanism. Those transients may arise as a result of a control command, load surge, supply voltage change, and so on. The duration of the transient processes, such as starting, braking, and transition from one speed to another must always be taken into account because it effects the drive performance [1–5].

The duration of the transient can be determined from the equation of motion (see section 1.2):

$$\left. \begin{array}{c} T_m - T_r = T_{dyn} = J_{eq} \cdot \dfrac{d\omega}{dt} \ \ [\text{Nm}] \\[2mm] \text{or} \\[2mm] T_m - T_r = T_{dyn} = \dfrac{(GD^2)_{eq}}{375} \cdot \dfrac{dn}{dt} \ \ [\text{Nm}] \end{array} \right\} \qquad (3.1)$$

where

T_m [Nm] is the driving torque developed by the motor,

T_r [Nm] is the resistive torque referred to the motor shaft,

T_{dyn} [Nm] is the dynamic torque at the motor shaft,

J_{eq} [Nm · s²] is the moment of inertia referred to the motor shaft,

doi:10.1088/978-0-7503-6104-0ch3

3-1

ω [s^{-1}] is the angular speed of the motor shaft,
t [s] is the time,
$(GD^2)_{eq}$ [Nm^2] is the flywheel moment referred to the motor shaft, and
n [rpm] is the speed of the motor shaft

Using equation (3.1), the transient time would be:

$$\left.\begin{aligned} t_{1,2} &= \int_{\omega_1}^{\omega_2} \frac{J_{eq}}{T_m - T_r} d\omega \ [\text{s}] \\ &\qquad\text{or} \\ t_{1,2} &= \int_{n_1}^{n_2} \frac{(GD^2)_{eq}}{375} \frac{1}{T_m - T_r} dn \ [\text{s}] \end{aligned}\right\} \tag{3.2}$$

where the speed varies between ω_1 and ω_2, or n_1 and n_2.

To perform the integration (equation (3.2)), it is necessary to know the torque versus speed behavior of the motor and of the mechanism.

Example 3.1 An electric motor drives a mechanism. The equivalent moment of inertia referred to the motor shaft is J_{eq} (figure E3.1(a)). The resistive torque referred to the motor shaft is constant at $T_r = T_n = $ Const. independent of the shaft speed (figure E3.1(b)). The starting torque of the motor is αT_n where α is a constant, $\alpha > 1$. Calculate the starting time of the drive from standstill to ω_1.

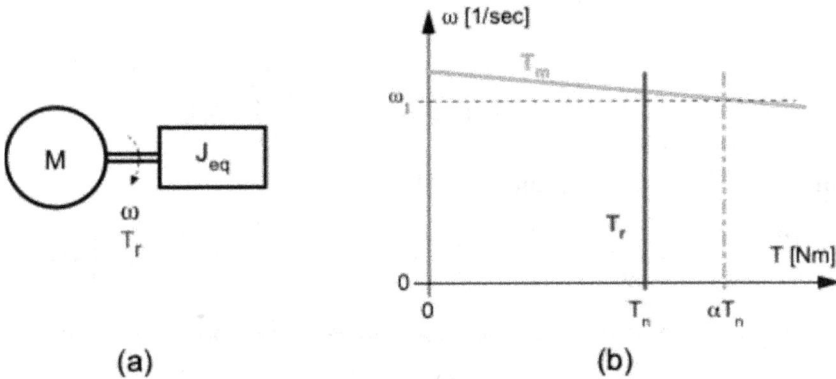

Figure E3.1. Transient time at constant starting torque. (a) Motor and load, and (b) speed versus torque curves.

Solution
Using equation (3.2), the starting time is:

$$t_{1,2} = \int_{\omega_1}^{\omega_2} \frac{J_{eq}}{T_m - T_r} d\omega = J_{eq} \int_0^{\omega_1} \frac{d\omega}{\alpha T_n - T_n} = \frac{J_{eq} \cdot \omega_1}{T_n(\alpha - 1)} \ [\text{s}]$$

Example 3.2 An electric motor drives a mechanism. The equivalent moment of inertia referred to the motor shaft is J_{eq} (figure E3.2(a)). The behavior of the dynamic torque T_{dyn} at the shaft of the motor is given by the linear curve (figure E3.2(b)).

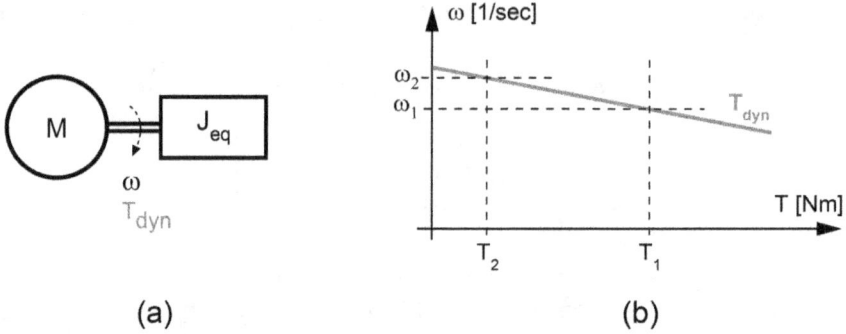

(a)   (b)

Figure E3.2. A linear curve of a dynamic torque. (a) Motor and load, and (b) speed versus torque curves.

Calculate the transition time it takes for the motor to vary its shaft speed from ω_1 to ω_2.

Solution

The equation of the linear curve of the dynamic torque is:

$$\frac{\omega - \omega_1}{\omega_2 - \omega_1} = \frac{T_{dyn} - T_1}{T_2 - T_1} \implies T_{dyn} = A \cdot \omega + B \ \ [\text{Nm}]$$

where,

$$A = \frac{T_2 - T_1}{\omega_2 - \omega_1} \ \text{and} \ B = \frac{T_1 \omega_2 - T_2 \omega_1}{\omega_2 - \omega_1}$$

The transient time between ω_1 and ω_2 is [1]:

$$t_{1,2} = J_{eq} \int_{\omega_1}^{\omega_2} \frac{d\omega}{A \cdot \omega + B} = J_{eq} \frac{1}{A} \cdot \ln \left[A \cdot \omega + B \right] \Big|_{\omega_1}^{\omega_2} = J_{eq} \frac{1}{A} \cdot \ln \frac{A \cdot \omega_2 + B}{A \cdot \omega_1 + B} \ \ [\text{s}]$$

Substituting the value of the coefficients A and B in the above equation:

$$t_{1,2} = J_{eq} \frac{\omega_2 - \omega_1}{T_2 - T_1} \cdot \ln \frac{\dfrac{T_2 \omega_2 - T_1 \omega_2}{\omega_2 - \omega_1} + \dfrac{T_1 \omega_2 - T_2 \omega_1}{\omega_2 - \omega_1}}{\dfrac{T_2 \omega_1 - T_1 \omega_1}{\omega_2 - \omega_1} + \dfrac{T_1 \omega_2 - T_2 \omega_1}{\omega_2 - \omega_1}}$$

$$= J_{eq} \frac{\omega_2 - \omega_1}{T_2 - T_1} \cdot \ln \frac{T_2 \omega_2 - T_1 \omega_2 + T_1 \omega_2 - T_2 \omega_1}{T_2 \omega_1 - T_1 \omega_1 + T_1 \omega_2 - T_2 \omega_1}$$

Finally: $t_{1,2} = J_{eq} \dfrac{\omega_2 - \omega_1}{T_2 - T_1} \cdot \ln \dfrac{T_2}{T_1} \ \ [sec]$

3-3

3.2 Number of revolutions during speed transients

In some industrial applications there is a need to know the position of the motor shaft or the number of revolutions when the speed varies, such as for an elevator reaching the floor level, or for a power drill reaching a given drilling depth, or for a tool machine performing a precise job, or for a door closing at a specified angle, and so forth.

Considering the motor shaft itself:

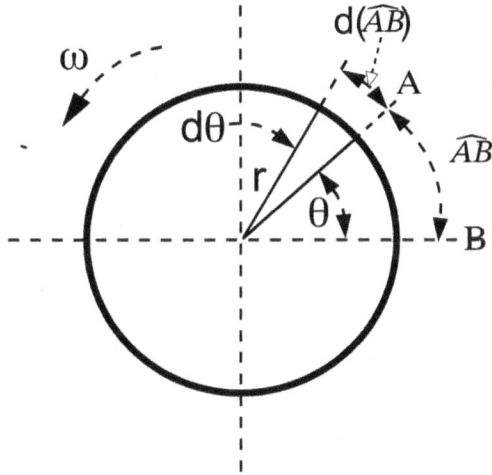

Figure 3.1. Angular displacement of the motor shaft.

A unit length of the arc $\widehat{AB}$ (figure 3.1) is:

$$d(\widehat{AB}) = r \cdot d\theta \qquad (3.3)$$

where
  $d(\widehat{AB})$ is a unit length of the arc $\widehat{AB}$
  r is the radius of the circle, and
  $d\theta$ is a unit of the angular displacement.

The length of the arc at one complete revolution is:

$$\widehat{AB}_C = \int_0^{2\pi} r \cdot d\theta = 2\pi \cdot r \qquad (3.4)$$

And the angular displacement of one complete revolution is:

$$\theta_C = \frac{\widehat{AB}_C}{r} = \frac{2\pi \cdot r}{r} = 2\pi \ [\text{rad}] \qquad (3.5)$$

For a multiple number of revolutions, N, the displacement angle would be:

$$\theta_N = 2\pi \cdot N \text{ [rad]} \tag{3.6}$$

Using the equation of motion (equation (3.1)):

$$T_m - T_r = J_{eq}\frac{d\omega}{dt} = J_{eq}\frac{d\omega}{d\theta} \cdot \frac{d\theta}{dt} = J_{eq} \cdot \omega\frac{d\omega}{d\theta} \text{ [Nm]} \tag{3.7}$$

where

T_m [Nm] is the driving torque developed by the motor,
T_r [Nm] is the resistive torque referred to the motor shaft,
J_{eq} [Nm s^2] is the moment of inertia referred to the motor shaft,
$\omega = d\theta/dt$ [s^{-1}] is the angular speed of the motor shaft,
t [s] is the time, and
θ [rad] is the angular displacement of the rotating shaft.

From equation (3.7), when the speed varies from ω_1 to ω_2, the angular displacement of the motor shaft would be:

$$\theta_N = J_{eq}\int_{\omega_1}^{\omega_2} \frac{\omega}{T_m - T_r}d\omega \text{ [rad]} \tag{3.8}$$

And the total number of revolutions is (equation (3.6)):

$$N = \frac{\theta_N}{2\pi} = \frac{J_{eq}}{2\pi}\int_{\omega_1}^{\omega_2} \frac{\omega}{T_m - T_r}d\omega \text{ [revolutions]} \tag{3.9}$$

Example 3.3 The electric motor of the drive in example 3.2 (figure E3.2) varies its speed from ω_1 to ω_2. Calculate the angular displacement and the number of revolutions of the motor shaft during that speed change.

Solution
From example 3.2, the dynamic torque at the motor shaft is:

$$T_{\text{dyn}} = T_m - T_r = A \cdot \omega + B \text{ [Nm]}$$

Using equation (3.8), the displacement angle would be [1]:

$$\theta_N = J_{eq}\int_{\omega_1}^{\omega_2} \frac{\omega \cdot d\omega}{A \cdot \omega + B} = J_{eq}\left(\frac{\omega_2 - \omega_1}{A} - \frac{B}{A^2} \cdot \ln\frac{A \cdot \omega_2 + B}{A \cdot \omega_1 + B}\right) \text{ [rad]}$$

Substituting the value of the coefficients A and B in the above equation:

$$\theta_N = J_{eq}\frac{\omega_2 - \omega_1}{T_2 - T_1}\left(\omega_2 - \omega_1 - \frac{T_1\omega_2 - T_2\omega_1}{T_2 - T_1} \cdot \ln\frac{T_2}{T_1}\right) \text{ [rad]}$$

And the number of revolutions of the motor shaft is (equation (3.6)):

$$N = \frac{\theta_N}{2\pi} = \frac{J_{eq}}{2\pi} \cdot \frac{\omega_2 - \omega_1}{T_2 - T_1}\left(\omega_2 - \omega_1 - \frac{T_1\omega_2 - T_2\omega_1}{T_2 - T_1} \ln \frac{T_2}{T_1}\right) \text{ [revolutions]}.$$

3.3 Use of finite increments method

The torque versus speed behavior of the motor and/or the mechanism can have a nonlinear nature. In that case, an exact solution of the transient time cannot be determined using algebraic methods, and an approximate solution must be applied. One approximate technique is called method of increments or finite increments method.

For a nonlinear case, the speed versus torque characteristic, $\omega = f(T)$, of the motor and its referred load can be given as a chart, such as a range of data points in a table or a graph. Using those charts, the finite increment method can be devised. The steps are:

(a) Determine the data points of the dynamic torque curve at the motor shaft:

$$T_{\text{dyn}}^{(i)} = T_m^{(i)} - T_r^{(i)} \text{ [Nm]} \tag{3.10}$$

where i is an increment number.

(b) At each increment, calculate a correlating average value of the dynamic torque in equation (3.10):

$$T_{d.\text{avg}}^{(i)} = T_{\text{dyn}}^{(i-1)} + \frac{T_{\text{dyn}}^{(i)} - T_{\text{dyn}}^{(i-1)}}{2} \text{ [Nm]} \tag{3.11}$$

where
$T_{d,\text{avg}}^{(i)}$ is the average value between two adjacent points on the dynamic torque curve,
$T_{\text{dyn}}^{(i)}$ is the dynamic torque at point (i), and
$T_{\text{dyn}}^{(i-1)}$ is the dynamic torque at the preceding point $(i - 1)$.
- That way, the actual dynamic torque curve becomes a piecewise constant function.

(c) Replace the infinitesimal values in the equation of motion (equation (3.1)) with finite increment values. That is:

$$\left.\begin{aligned} d\omega_i &\cong \Delta\omega_i = \omega_i - \omega_{i-1} \ \ [\text{s}^{-1}] \\ dn_i &\cong \Delta n_i = n_i - n_{i-1} \ [\text{rpm}] \\ dt &\cong \Delta t_i = t_i - t_{i-1} \ [\text{s}] \end{aligned}\right\} \tag{3.12}$$

where
ω_i is the radial shaft speed at point (i),
ω_{i-1} is the preceding radial shaft speed at point $(i - 1)$,
n_i is number of revolutions at point (i),
n_{i-1} is the preceding number of revolutions at point $(i - 1)$,
$\Delta\omega_i$ is an angular speed increment between two adjacent points,
Δn_i is a number of revolutions between two adjacent points, and
Δt_i is a time between two adjacent points.

- That way, at each increment i, the equation of motion is:

$$T_{d,avg}^{(i)} = J_{eq} \frac{\Delta\omega_i}{\Delta t_i} \quad [\text{Nm}]$$

$$\text{or:} \quad T_{d,avg}^{(i)} = \frac{(GD^2)_{eq}}{375} \frac{\Delta n_i}{\Delta t_i} \quad [\text{Nm}]$$

(3.13)

where

J_{eq} is the moment of inertia referred to the motor shaft and

$(GD^2)_{eq}$ is the flywheel moment referred to the motor shaft.

(d) Calculate the total transient time t_n using the equation of motion (equation (3.13)) at each increment:

$$t_n = J_{eq} \sum_{i=1,\ 2.\ 3...}^{n} \frac{\Delta\omega_i}{T_{d,avg}^{(i)}} \quad [\text{s}]$$

$$\text{or:} \quad t_n = \frac{(GD^2)_{eq}}{375} \sum_{i=1,\ 2.\ 3...}^{n} \frac{\Delta n_i}{T_{d,avg}^{(i)}} \quad [\text{s}]$$

(3.14)

where n is the final increment.

Example 3.4 An electric motor drives a mechanism (figure E3.4.1(a)). The nonlinear nature of the speed–torque curve, T_m, of the motor is presented in figure E3.4.1(b), and its data points are given in table E3.4.1. The nonlinear nature of the referred resistive torque, T_r, is also presented in the figure and its data points are given in the table. The equivalent flywheel moment referred to the motor shaft is $(GD^2)_{eq} = 200 \ [\text{Nm}^2]$.

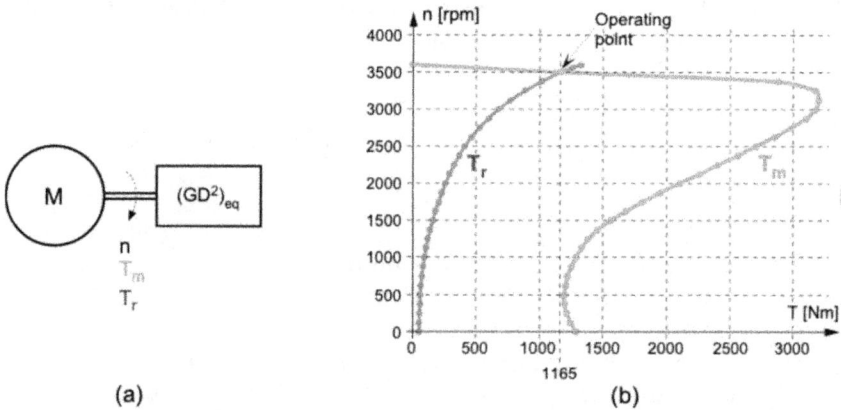

Figure E3.4.1. An electromechanical drive displaying nonlinear torque curves. (a) Motor and load, and (b) speed–torque curves.

Table E3.4.1. Data points for the speed–torque curves displayed in figure E3.4.1b, where T_m and T_r are given in [Nm], and the speed n in [rpm].

	1	2	3	4	5	6	7	8	9	10	11	12	13	14	15	16
T_m	1290	1248	1220	1198	1200	1210	1225	1255	1290	1335	1386	1460	1560	1680	1810	1950
T_r	60	52	54	57	60	65	72	81	92	105	120	137	158	180	205	230
n	0	125	250	375	500	625	750	875	1000	1125	1250	1375	1500	1625	1750	1875

	17	18	19	20	21	22	23	24	25	26	27	28	29	30	31
T_m	2100	2255	2400	2560	2700	2850	2980	3100	3180	3200	3180	2880	1165	500	0
T_r	258	290	326	365	410	463	528	600	682	778	888	1020	1165	1260	1330
n	2000	2125	2250	2375	2500	2625	2750	2875	3000	3125	3250	3375	3500	3560	3600

Calculate the starting time of the drive. That is, the time it takes to reach the operating point speed of $n = 3500$ [rpm] from a standstill $n = 0$ [rpm].

Solution

One way of applying the finite increments method would be by making use of a straightforward computer program such as a spreadsheet editor.

Using the data points in table E3.4.1, the solution can be obtained by using the procedure set forth in section 3.3. The calculation results are listed in table E3.4.2.

Examples of the four steps are:

(a) The dynamic torque increment $T_{dyn}^{(i)}$ is (equation (3.10))

$$T_{dyn}^{(i)} = T_m^{(i)} - T_r^{(i)}$$

$$\left(\begin{array}{l} \textbf{Example:} \ \ \text{Point \#10 in table E3.4.2} \\ T_{dyn}^{(i=10)} = 1335 - 105 = 1230 \ [\text{Nm}] \end{array} \right)$$

(b) The average torque increment $T_{d.\text{avg}}^{(i)}$ is (equation (3.11)):

$$T_{d.\text{avg}}^{(i)} = T_{dyn}^{(i-1)} + \frac{T_{dyn}^{(i)} - T_{dyn}^{(i-1)}}{2}$$

$$\left(\begin{array}{c} \textbf{Example:} \ \ \text{Point \#10 in table E3.4.2} \\ T_{d.\text{avg}}^{(i=10)} = 1198 + \frac{1230 - 1198}{2} = 1214 \ [\text{Nm}] \end{array} \right)$$

(c) The speed increment Δn_i is (equation (3.12)):

$$\Delta n_i = n_i - n_{i-1}$$

$$\left(\begin{array}{l} \textbf{Example:} \ \ \text{Point \#10 in table E3.4.2} \\ \Delta n_{i=10} = 1125 - 1000 = 125 \ [\text{rpm}] \end{array} \right)$$

(d) And the time increment t_i is (equation (3.14)):

$$t_i = t_{i-1} + \frac{(GD^2)_{eq}}{375} \frac{\Delta n_i}{T_{d.\text{avg}}^{(i)}}$$

$$\left(\begin{array}{l} \textbf{Example:} \ \ \text{Point \#10 in table E3.4.2} \\ t_{i=10} = 0.46 + \frac{200}{375} \frac{125}{1214} = 0.51 \ [\text{s}] \end{array} \right)$$

Finally: The total startup time of the drive is (Point \#29 in table E3.4.2):

$$t_n = 1.23 \ [\text{s}]$$

Table E3.4.2. Given data and calculation results (from a spreadsheet editor).

	Given data points			Calculation results			
	T_m [Nm]	T_r [Nm]	n [rpm]	$T_{dyn}^{(i)}$ [Nm]	$T_{d.avg}^{(i)}$ [Nm]	Δn_i [rpm]	t_i [s]
1	1290	50	0	1240			0.00
2	1248	52	125	1196	1218.0	125	0.05
3	1220	54	250	1166	1181.0	125	0.11
4	1198	57	375	1141	1153.5	125	0.17
5	1200	60	500	1140	1140.5	125	0.23
6	1210	65	625	1145	1142.5	125	0.29
7	1225	72	750	1153	1149.0	125	0.34
8	1255	81	875	1174	1163.5	125	0.40
9	1290	92	1000	1198	1186.0	125	0.46
10	1335	105	1125	1230	1214.0	125	0.51
11	1386	120	1250	1266	1248.0	125	0.57
12	1460	137	1375	1323	1294.5	125	0.62
13	1560	158	1500	1402	1362.5	125	0.67
14	1680	180	1625	1500	1451.0	125	0.71
15	1810	205	1750	1605	1552.5	125	0.75
16	1950	230	1875	1720	1662.5	125	0.80
17	2100	258	2000	1842	1781.0	125	0.83
18	2255	290	2125	1965	1903.5	125	0.87
19	2400	326	2250	2074	2019.5	125	0.90
20	2560	365	2375	2195	2134.5	125	0.93
21	2700	410	2500	2290	2242.5	125	0.96
22	2850	463	2625	2387	2338.5	125	0.99
23	2980	528	2750	2452	2419.5	125	1.02
24	3100	600	2875	2500	2476.0	125	1.04
25	3180	682	3000	2498	2499.0	125	1.07
26	3200	778	3125	2422	2460.0	125	1.10
27	3180	888	3250	2292	2357.0	125	1.13
28	2880	1020	3375	1860	2076.0	125	1.16
29	1165	1165	3500	0	930.0	125	1.23
30	500	1260	3560				
31	0	1330	3600				

Four charts in table E3.4.2: $n = f_1(T_m)$, $n = f_2(T_r)$, $n = f_3(T_{dyn})$, and $n = f_4(t)$, are presented graphically in figure E3.4.2.

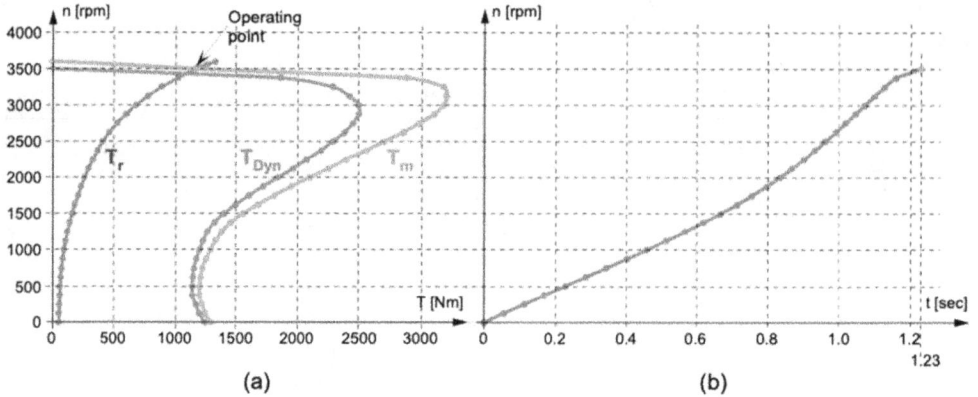

Figure E3.4.2. Startup time of the electromechanical drive. (a) Referred resistive torque, dynamic torque, motor torque, and (b) speed-time curve.

3.4 Problems

1. An electric motor has a flywheel moment $(GD^2)_m = 0.4$ [Nm²], and its rated speed is $n_n = 1,430$ [rpm]. At rated speed, the motor drives a cart at a linear velocity of $v = 1.5$ [m s⁻¹]. During startup, the equivalent dynamic torque at the motor shaft is $T_{dyn} = 7$ [Nm]. The resistive torque referred to the motor shaft is $T_r = 4$ [Nm], and the referred (equivalent) flywheel moment of the mechanism is $(GD^2)_{eq} = 3.1$ [Nm²].

   Calculate the distance the cart moves during startup.

2. Calculate the starting time of a drive with the following parameters:
   The motor speed is: $\omega_m = 88.9 - 0.72 \cdot T$ [s⁻¹]
   Resistive torque referred to the motor shaft is: $T_r = 5.21 + 0.58 \cdot \omega$ [Nm]
   The equivalent moment of inertia referred to the motor shaft is:
   $J_{eq} = 0.12$ [kg m²]

   $$\text{Reminder: } y = \int \frac{dx}{a + bx} = \frac{1}{b} \ln(a + bx)$$

3. An electric motor drives a winding drum through a gearbox (figure P3.3). The moment of inertia of the motor is $J_m = 0.2$ [kg$_m$m²]. The moment of

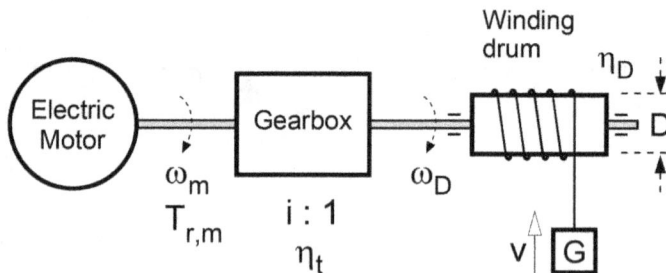

Figure P3.3. An electric motor drives a winding drum via a gearbox.

inertia of the gearbox, referred to its low-speed shaft, is $J_G = 15 \, [\text{kg}_\text{m}\text{m}^2]$, its gear ratio is $i = 10$, and its efficiency is $\eta_t = 0.9$.

The diameter of the drum is $D = 0.4 \, [\text{m}]$, its moment of inertia is $J_D = 15 \, [\text{kg}_\text{m}\text{m}^2]$, and its efficiency is $\eta_D = 0.95$. The load weighs $G = 500 \, [\text{kg}]$. During startup, from standstill to full velocity of $= 1 \, [\text{m s}^{-1}]$, the load accelerates at $a = 0.5 \, [\text{m s}^{-1}]$.

Calculate the starting torque required at the motor shaft.

4. An electric motor drives a winding drum through a gearbox (as in figure P3.3). The motor has a moment of inertia $J_m = 0.5 \, [\text{kg} \cdot \text{m}^2]$. The gearbox has a gear ratio $i = 10$, a moment of inertia $J_g = 0.15 \, [\text{kg} \cdot \text{m}^2]$ referred to its high-speed shaft, and an efficiency $\eta_t = 0.95$. The winding drum has a diameter $D = 0.4 \, [\text{m}]$ a moment of inertia $J_D = 10 \, [\text{kg} \cdot \text{m}^2]$, and an efficiency $\eta_D = 0.95$. The load itself weighs $G = 500 \, [\text{kg}]$.

When the load is elevated, the drive starts up in two stages (figure P3.4): in the first stage, from 0 to ω_1, the starting torque develops by the motor is constant at $T_{st} = 200 \, [\text{Nm}]$. In the second stage, from ω_1 to the operating point at ω_2, the starting torque of the motor follows the curve $\omega = 172 - 0.13 \, T$.

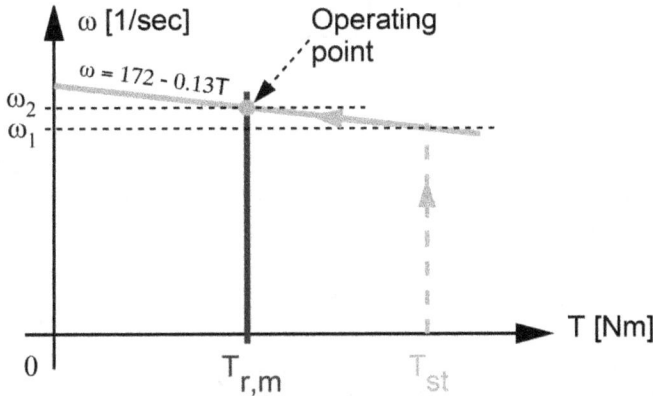

Figure P3.4. Two-stage startup.

Calculate:
(a) The speed of the motor shaft at the operating point.
(b) The equivalent moment of inertia referred to the motor shaft.
(c) The require power of the motor.
(d) The total startup time of the drive from $\omega = 0$ to $\omega = \omega_2$.

$$\text{Reminder: } y = \int \frac{dx}{a + bx} = \frac{1}{b} \ln(a + bx)$$

5. The three-speed hoist-machine (figure P3.5) was invented by Brunelleschi at the beginning of the 15th century (three axles at different speeds). At that time, a horse was the prime mover. Today, we might use an electric motor.

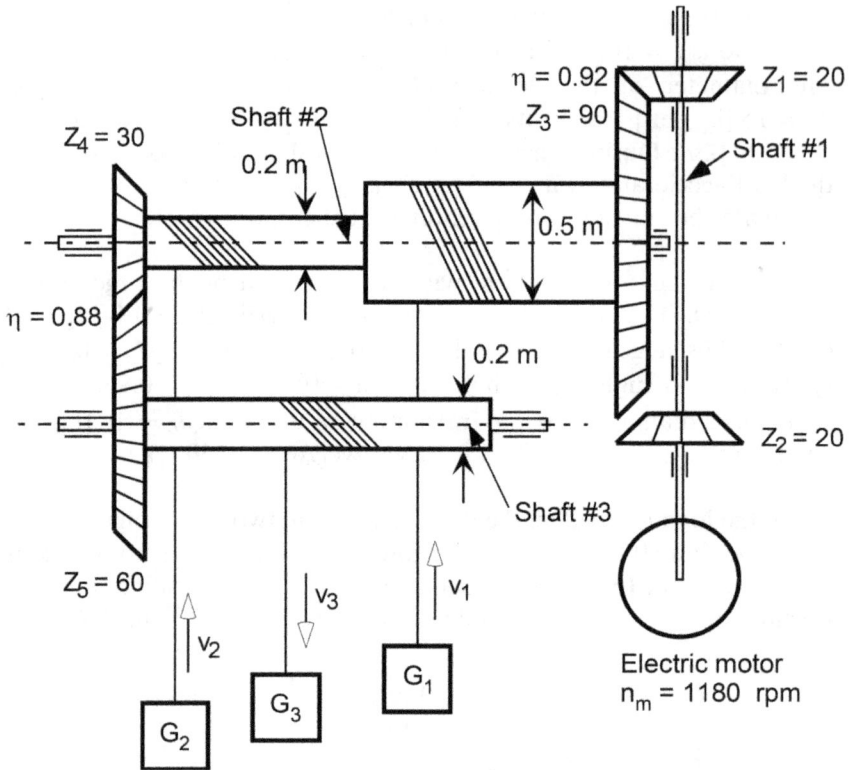

Figure P3.5. Three-speed hoist invented by Brunelleschi (1420).

Shaft #1 with its two cogwheels Z_1 and Z_2 is attached to the electric motor and can move up and down to engage with cogwheel Z_3 for a forward or a reverse operation. When Z_1 is engaged, as shown in the diagram (figure P3.3), the motor rotates at a rated speed of $n_m = 1180$ [rpm].

Given:

(a) The flywheel moment of Shaft #1, which includes the motor and the two cogwheels Z_1 and Z_2, is $(GD^2)_1 = 0.2$ [Nm²].

(b) The flywheel moment of Shaft #2, which includes the two winding drums of diameters $D_{2,1} = 0.5$ [m] and $D_{2,2} = 0.2$ [m] and the two cogwheels Z_3 and Z_4, is $(GD^2)_2 = 6.2$ [Nm²].

(c) The flywheel moment of Shaft #3, which includes the third winding drum of $D_3 = 0.2$ [m] diameter and cogwheel Z_5, is $(GD^2)_3 = 4.4$ [Nm²].

(d) The three loads are: $G_1 = 100$ [N] that moves upward at a velocity v_1; $G_2 = 400$ [N] that also moves upward but at v_2; and $G_3 = 1000$ [N] that moves downward at v_3.

Calculate the starting time of the system, where the motor torque at start is constant at twice the referred resistive torque at the motor shaft.

6. A single electric motor drives a train at a linear speed of $v = 67.1$ [mph]. The train consists of three cars (figure P3.6a). The motor, weighs $G_m = 300$ [kg] and has a flywheel moment $(GD^2)_m = 200$ [Nm²]. The motor is connected to two front wheels (figure P3.6b) through a gearbox that weighs $G_g = 100$ [kg] and has a flywheel moment $(GD^2)_g = 100$ [Nm²] (referred to its high-speed shaft). The gearbox has a gear ratio $i = 6$, and its efficiency is $\eta = 0.9$. The wheels of all cars have the same diameter $D = 1$ [m], and each wheel weighs $G_w = 100$ [kg] (including shafts). The wheel-rail friction of each wheel is $F = 2000$ [N]. The weight of each car itself, not including the wheels and the motor and the gearbox, is $G_C = 2000$ [kg]. The moment of inertia of each wheel is given by $J_w = M_w \rho_w^2$, where M_w is the mass of the wheel and $\rho_w = D/2\sqrt{2}$ [m] is its radius of gyration.

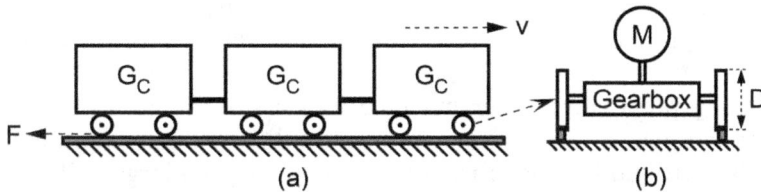

Figure P3.6. An electric train. (a) three cars train, (b) motor and gearbox at the front wheels.

The motor is disconnected from its power source and the train decelerates. System braking is caused only by friction at the wheels. Calculate the time it takes for the train to come to a complete stop.

7. A motorized escalator has a circulating metallic belt with folding stairs (figure P3.7). The dimensions of each unfolded stair are: width $W = 0.3$ [m]

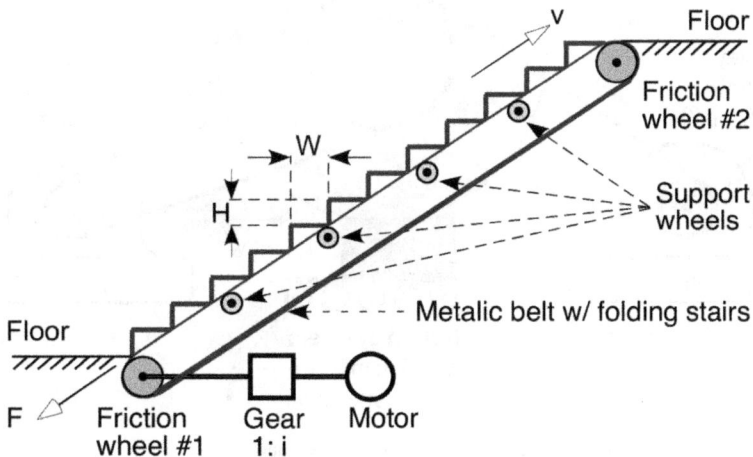

Figure P3.7. A motorized escalator.

and height $H = 0.2$ [m]. The 12 unfolded stairs are designed to convey 7200 people per hour, where each stair is intended for 2 people at 200 [lb] each.

The total weight of the moving belt, which includes the folded and unfolded stairs, is $G_b = 2000$ [kg]. Each support wheel (total of four) has a diameter of $D_s = 0.2$ [m], and a moment of inertia of $J_s = 50$ [Nm · s^2]. Each friction wheel (total of two) has a diameter of $D_f = 0.5$ [m], and a moment of inertia of $J_f = 500$ [kg · m^2]. The motor itself has a moment of inertia od $J_m = 0.3$ [Nm · s^2]. The gearbox moment of inertia is $J_g = 0.6$ [kg · m^2], its efficiency is $\eta = 0.95$, and it has a gear ratio of $i = 60$. At maximum load, the friction force of the metallic belt itself is $F = 10^4$ [N].
Calculate:

   (1) the rated power [kW] of the motor, and
   (2) the startup time of the drive at full capacity, where the motor startup torque is constant at twice the referred resistive torque at its shaft.

8. A baling press car crusher machine uses a solid metal drum to vertically crush scrap cars, refrigerators, and air-conditioners for recycling. The drum revolves around its axis of rotation with the aid of a large pulley attached to its shaft at an efficiency of $\eta_D = 0.9$. Two identical large electric motors employing a two-belt drive system, efficiency $\eta_b = 0.85$, are used to rotate the large pulley. The scrap car is placed on a heavy steel plate that is supported by 12 rolling cylinders (figure P3.8).

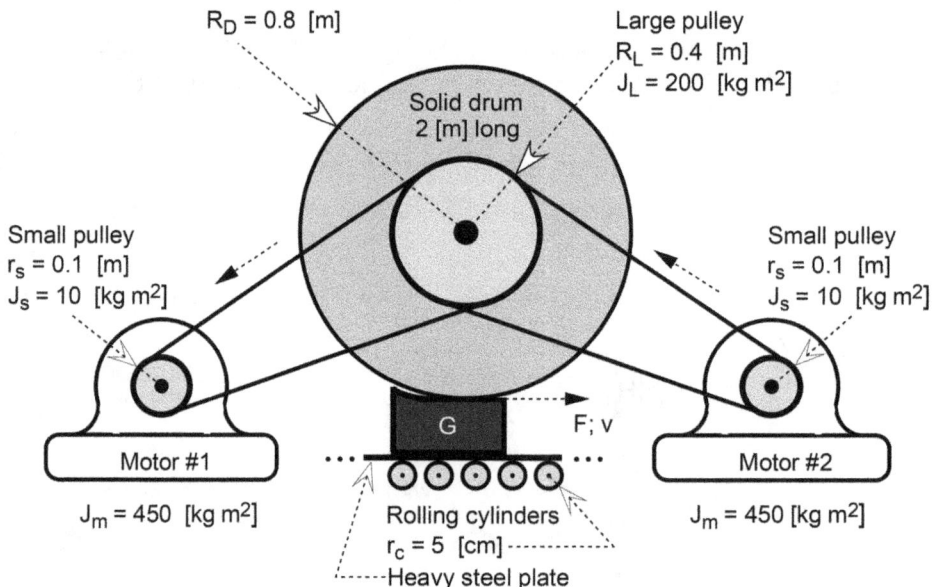

Figure P3.8. A car crusher machine.

The total load, which includes the scrap car, the steel plate and the rolling cylinders, moves linearly at a velocity of $v = 14.3$ [m s^{-1}]. The rated force required to move that load is $F = 35\,000$ [kg$_f$].

- The moment of inertia of the solid metal drum is $J_D = M(R_D^2/2)$ [kg m^2], where M [kg$_m$] is its mass and R_D [m] is its radius. The density of the drum material (steel) is $\gamma = 7.9 \cdot 10^3$ [kg m^{-3}].
- The rated weight of the scrap car is $G = 4000$ [kg].
- The weight of the steel plate is $G_P = 750$ [kg].
- The weight of each rolling cylinder is $G_c = 125$ [kg], and its moment of inertia is $J_c = 2$ [kg m^2].
- Additional data is given in the figure itself.

Calculate:
- (a) The rated power of each of the two motors.
- (b) Assume a constant starting torque of each motor to be twice the resistive torque at its shaft, and calculate the starting time of the drive first at no-load, and then with the load.

9. A two-wheel electric scooter that weighs $G_S = 52$ [lb] (including wheels, motor, chain, and battery pack) is designed to carry a person that weighs $G_P = 220$ [lb]. Its back drive-wheel is propelled by a chain drive-motor (figure P3.9)).

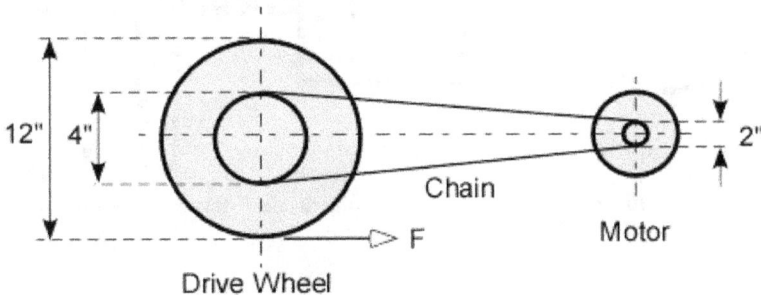

Figure P3.9. Electric scooter mechanism.

- Maximum speed at full load is 15 [mph].
- The friction force at full load at each wheel is: $F = 2.5$ [kg$_f$].
- The chain length is $L_C = 1.2$ [m], and its weight is $G_C = 15$ [kg m^{-1}].
- The flywheel moment of the motor, including its shaft is $(GD^2)_m = 50$ [Nm^2].

- The flywheel moment of the drive wheel, including the shaft and chain-wheel is: $(GD^2)_D = 100$ [Nm²].
- The flywheel moment of the second (front) wheel is: $(GD^2)_F = 80$ [Nm²].
- The efficiency of the drive itself is: $\eta = 0.9$.

Calculate:
 (a) The required power of the motor.
 (b) At full load, assume that the motor develops a constant starting torque ten (10) times the resistive torque at its shaft, and calculate the starting time of the scooter from standstill to full speed.

10. A four-wheel electric wheelchair is driven by a single motor connected to its left back wheel via a gearbox. The block-diagram shows the mechanical connection of the motor, the gearbox, and its back left wheel (figure P3.10).

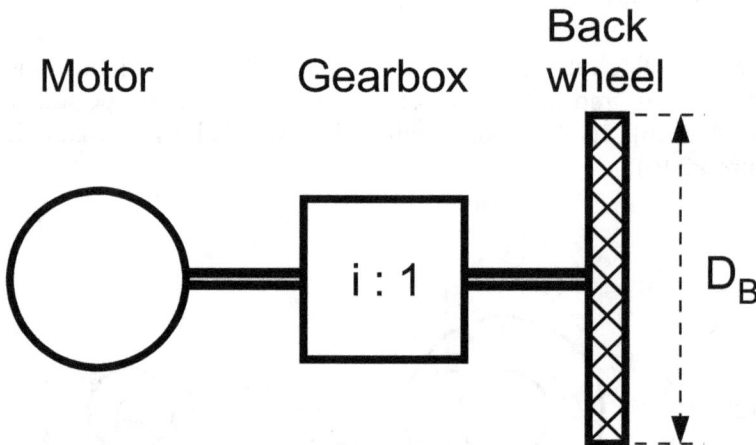

Figure P3.10. Block diagram of the electromechanical system of a wheelchair.

- Total weight of the wheelchair is $G_C = 25$ [kg].
- Total capacity (full load, which is the weight of a person) is $G_P = 150$ [kg].
- Front wheel diameter is $D_F = 20$ [cm].
- Rear (back) wheel diameter is $D_B = 25$ [cm].
- The friction force, wheel to track (road), at full load of each of the front wheels is $F_F = 2$ [kg$_f$].
- The friction force, wheel to track (road), at full load of each of the back wheels is $F_B = 3$ [kg$_f$].
- The flywheel moment of the motor is $(GD^2)_m = 1.4$ [Nm²].

- The flywheel moment of the gear (related to its high-speed shaft) is $(GD^2)_g = 0.54$ [Nm^2].
- The flywheel moment of each of the front wheels is $(GD^2)_F = 31$ [Nm^2].
- The flywheel moment of each of the back wheels is $(GD^2)_B = 56$ [Nm^2].
- The gear ratio is $i = 10$.
- The efficiency of the gearbox is $\eta = 0.9$.
- The maximum velocity of the wheelchair at full load is $v = 8$ [kmh].

Calculate:
  (a) the motor power at full load and at rated (maximum) speed.
  (b) the starting time (from standstill to max velocity) of the wheelchair while carrying its full load. Assume that motor starting torque is constant at 10 times its referred resistive torque.

11. An aerial cable car suspended on two track cables is moving downhill at a constant speed of $v = 9$ [m s^{-1}]. The car is hooked to a carrier basket that is pulled by a haulage rope suspended between and wound around two friction wheels (figure P3.11). The basket itself has four wheels that move

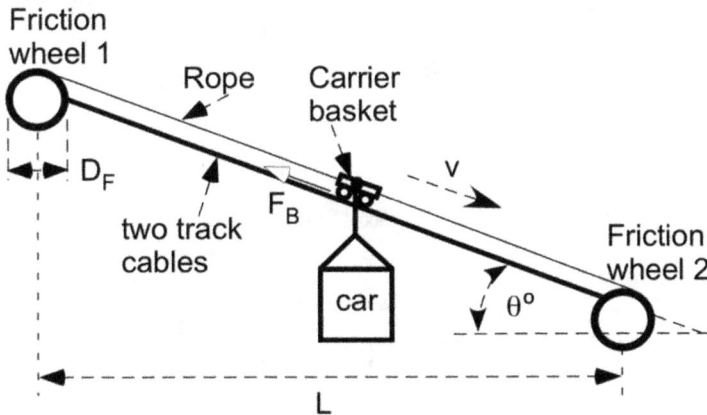

Figure P3.11. An aerial cable car.

on those track cables. An electric motor coupled to a gearbox (not seen in the figure) drives the carrier basket via its front wheels. When a braking signal is given, the motor stops the basket in $t_B = 5$ [s].

- The flywheel moment of each friction wheel is: $[GD^2]_F = 1000$ [Nm^2].
- The diameter of each friction wheel is: $D_F = 1$ [m].
- The weight of the car and its hook is: $G_C = 200$ [kg].
- The carrier basket weighs $G_B = 25$ [kg].
- Each of the four wheels has a friction force with the track cables of $F_B = 30$ [kg$_f$].

- Each wheel of the carrier basket has a $D_S = 0.2$ [m] diameter and a flywheel moment of $[GD^2]_S = 20$ [Nm²].
- Maximum weight of the car-load itself is: $G_L = 400$ [kg].
- The slop angle is $\theta = 30°$.
- The horizontal distance between the two friction wheels is: $L = 2$ [km].
- The gearbox efficiency is: $\eta = 0.95$.
- The gear ratio is: $i = 10$ (motor faster than the friction wheel).
- The rope weight is: $G_r = 0.1$ [kg m^{-1}].
- The combine flywheel moment of motor and gearbox is: $[GD^2]_m = 40$ [Nm²].

Assume that the motor develops a constant braking torque, and calculate:
  (a) The speed of the motor shaft as the car travels at a linear velocity v.
  (b) The required braking torque of the motor to stop the cable car within 5 [s].
  (c) The distance the car travels during the braking condition in (b).

12. An electric motor drives a mechanism (figure P3.12). The data points of the nonlinear speed–torque curve, T_m, of the motor is given in table P12 (same as in example 3.4). The resistive torque referred to the motor shaft is constant, $T_r = 1165$ [Nm]. The equivalent flywheel moment referred to the motor shaft is $(GD^2)_{eq} = 200$ [Nm²].

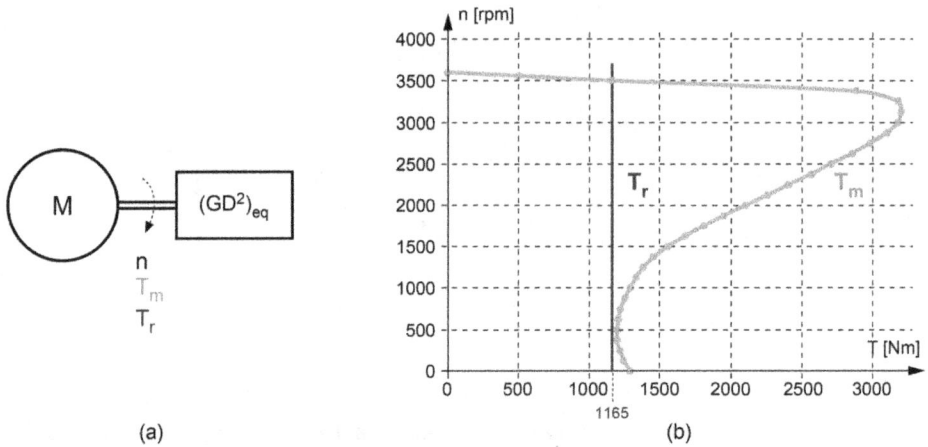

Figure P3.12. An electromechanical drive having a nonlinear motor torque. (a) Motor and load, and (b) speed–torque curves.

Calculate the starting time of the drive. That is, the time it takes to reach the operating point speed of $n = 3500$ [rpm] from a standstill $n = 0$ [rpm]. Also, draw the characteristic of motor speed versus time, $n = f(t)$.

Table P12. Data points for the motor speed–torque curve displayed in figure P3.12(b), where T_m is given in [Nm], and the speed n in [rpm].

	1	2	3	4	5	6	7	8	9	10	11	12	13	14	15	16
T_m	1290	1248	1220	1198	1200	1210	1225	1255	1290	1335	1386	1460	1560	1680	1810	1950
n	0	125	250	375	500	625	750	875	1000	1125	1250	1375	1500	1625	1750	1875

	17	18	19	20	21	22	23	24	25	26	27	28	29	30	31
T_m	2100	2255	2400	2560	2700	2850	2980	3100	3180	3200	3180	2880	1165	500	0
n	2000	2125	2250	2375	2500	2625	2750	2875	3000	3125	3250	3375	3500	3560	3600

References

[1] Mohan N 2001 *Electric Drives: An Integrative Approach* (MNPERE Publisher)
[2] Dubey G K 2001 *Fundamentals of Electrical Drives* (Alpha Science International Ltd)
[3] Nasar S A and Unnewehr L E 1983 *Electromechanical and Electric Machines* (New York: Wiley)
[4] Meyer R A (ed) 2002 *Encyclopedia of Physical Science and Technology* **vol 5** 3rd edn (New York: Academic)
[5] Chilikin M 1976 *Electric Drive* (MIR Publishers)

Part II

The electrical part

The electric motor is the prime mover of the electromechanical drive system. The motor is required to address the needs of the working machine in steady state operations, in controlling the mechanism speed, and in braking modes of operation. This part addresses three motor types: the shunt-wound connected DC motor, the induction motor, and the synchronous motor operating as a brushless DC motor.

IOP Publishing

Fundamentals of Electromechanical Drives

Zivan Zabar

Chapter 4

The direct current, DC, motor

The classical direct current, DC, motor is a fundamental electric machine. Today, it has limited use mainly due to the wear of its mechanical rectifier, but also for its reduced efficiency and increased weight (and increased capital cost). Still, its speed–torque characteristics and its braking mode capabilities are the criteria for most electromechanical drives. This chapter addresses briefly its principle of operation, its power and torque relationships, its braking modes of operation, and its speed control techniques.

4.1 Relevant fundamentals of electromagnetics—an overview

A magnetic circuit consists of a structure composed for the most part of high-permeability magnetic materials. The presence of those high-permeability materials causes the magnetic flux to be confined to the paths defined by the structure (the same as electric currents are confined to the conductors of an electric circuit). The magnetic flux lines themselves form closed loops [1–3].

- *Magnetic flux* (ϕ) is measured by the total number of magnetic flux lines passing through a specified area in a magnetic field. Assuming a uniform magnetic flux density, the total magnetic flux is:

$$\phi = \int B \cdot dA \ [\text{V} \cdot \text{s}] \tag{4.1}$$

  where
    ϕ [Weber $\equiv$ V $\cdot$ s] is the magnetic flux,
    B [Tesla $\equiv$ V $\cdot$s m^{-2}] is the magnetic flux density, and
    dA [m^2] is a unit area.
- *Ampere's circuital law* states that the line integral of the magnetic field surrounding closed-loop equals the number of times the algebraic sum of currents passing through the loop (equation (4.1)):

doi:10.1088/978-0-7503-6104-0ch4

$$\oint H \cdot dl = \sum N \cdot i \ \text{[A]} \tag{4.2}$$

where

H [A m^{-1}] is the magnetic field intensity,
dl [m] is a unit length,
N is the number of current-carrying conductors, and
i [A] is the electric current.

Example 4.1 Calculate the magnetic intensity around an infinite straight wire carrying current.

Solution

The magnetic field produced by an infinite current-carrying conductor encircles the conductor and lies in a plane perpendicular to the conductor (figure E4.1). The direction of the magnetic flux lines is determined by an empirical rule, called the curl right-hand rule.

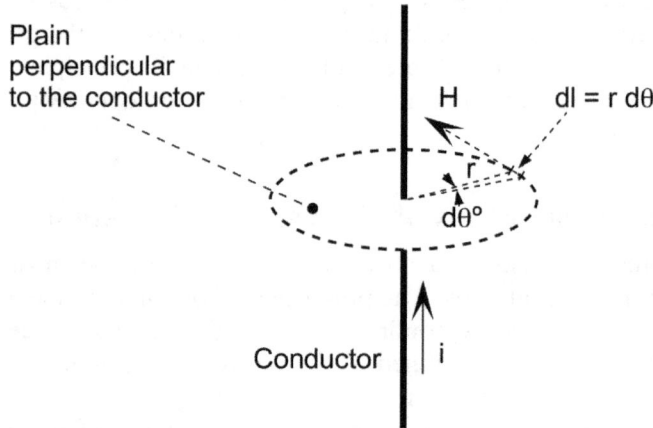

Figure E4.1. Magnetic field around an infinite long current-carrying conductor.

The curl right-hand rule states: curl the conductor with the right-hand fingers such that the thumb points in the direction of the electric current. The fingers indicate the direction of the magnetic field.

The magnetic intensity at a distance r away from the conductor is (equation (4.2)):

$$\int_0^{2\pi} H \cdot rd\theta = i \implies H = \frac{i}{2\pi \cdot r} \ \text{[A m}^{-1}\text{]}$$

- *Permeability* (μ) is the ability of a medium to allow magnetic flux to pass through it. It is measured by the ratio of the magnetic flux density (B) over the intensity (H) of the magnetic field:

$$\mu = \frac{dB}{dH} \left[\frac{\dfrac{V \cdot s}{m^2}}{\dfrac{A}{m}} = \frac{V \cdot s}{A \cdot m} = \frac{\Omega \cdot s}{m} \right] \tag{4.3}$$

Vacuum permeability is considered as the *magnetic constant*: $\mu_0 = 4\pi \cdot 10^{-7}$ [$\Omega \cdot$s m^{-1}].

- Electromagnetic induction is the generation of electromotive force (EMF) in a conductor when a change in the surrounding magnetic flux occurs. It is quantified by Faraday's Law:

$$e = -N\frac{d\varnothing}{dt} \text{ [V]} \tag{4.4}$$

where

e [V] is the induced voltage,
N is the number of conductors,
$d\phi$ [V $\cdot$ s] is a unit magnetic flux, and
dt [s] is a unit of time.

The negative sign (equation (4.4)) indicates that in a closed loop, the induced current generates magnetic flux that opposes the flux change.

4.2 Principle of operation

A direct current, DC, machine is an electrical rotating machine that can operate as a generator supplying DC voltage to an external circuit, or can operate as a motor supplied by a DC voltage source and provides torque at its shaft [1–3]. The machine consists of two main components: a stationary part and a rotating part. To allow rotation, there is a narrow airgap, 1–2 mm, between the two main parts (figure 4.1).

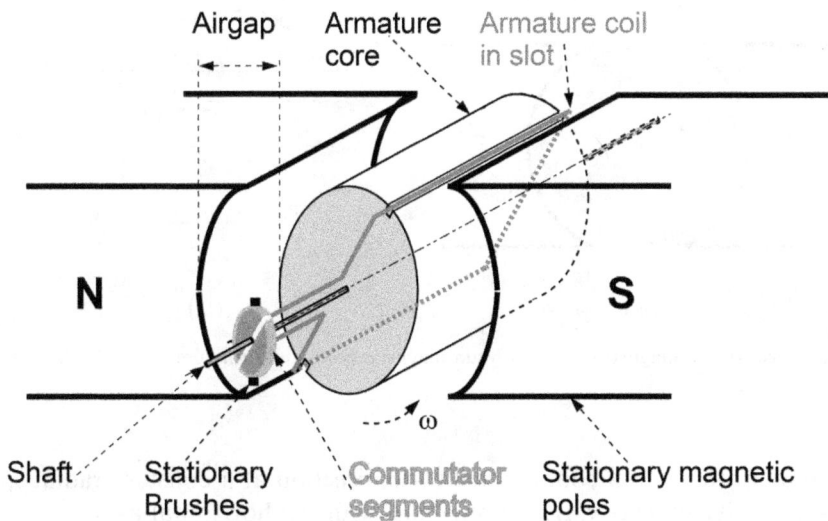

Figure 4.1. 3D view of the main parts of a DC machine.

- *The stationary part* (stator) holds the electromagnetic poles that produce the main magnetic field in the airgap of the machine. It also holds the brushes that provide galvanic connection between the rotating commutator and the external electric circuit.
- *The rotating part* (rotor), also terms the armature, spins around the machine shaft at a speed ω [s^{-1}]. It consists of laminated ferromagnetic materials. The armature conductors (coils) are evenly distributed in slots around its lateral surface. The coil terminals are connected to rotating commutator segments. Stationary brushes distribute the armature current to an external circuit. In figure 4.1, only one armature coil is shown. The terminals of that coil are connected to the two detached (insulated) copper segments of the commutator.

Assuming that a prime mover drives the armature (rotor) in a counter-clockwise direction at a radial speed of ω, and that the stationary brushes are connected to an external circuit (figure 4.1). The poles produce a magnetic flux ϕ that is perpendicular to the lateral surface area of the rotor. Voltage (EMF) is induced in the rotating coil, and current flows to the external circuit. The direction of the induced current in the conductors is determined by the open right-hand rule.

The open right-hand rule states: hold the open right hand such that the fingers show the direction of the magnetic flux and the driving force penetrates the palm of the hand. The extended thumb indicates the direction of the induced current.

Using the above rule, the direction of the induced current in the conductor that moves along the North magnetic pole is out of the page (figure 4.2(a)), and in the one that moves along the South pole is into the page. As a result, the current in each conductor reverses direction twice during one complete revolution; i.e., an alternating current (AC) is generated in the rotating coil.

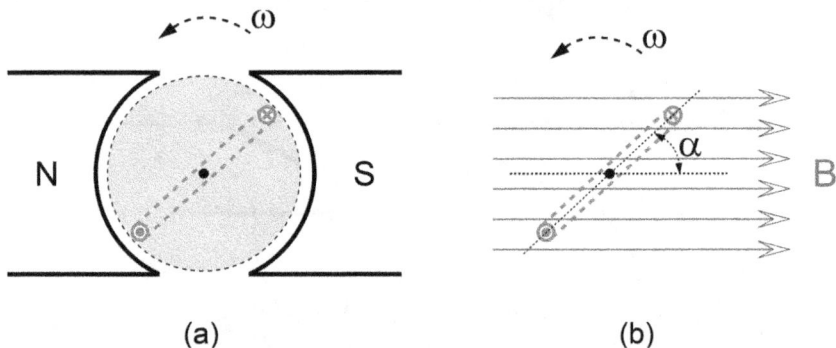

(a)   (b)

Figure 4.2. Effect of the magnetic field. (a) Main magnetic poles and armature, and (b) effect of the flux change.

As the coil rotates in a counter-clockwise direction at a constant radial speed ω (figure 4.2(b)), its angle α varies in reference with the horizontal axis:

$$\alpha = \omega \cdot t \ [\text{rad}] \tag{4.5}$$

Assuming a uniform magnetic flux density B along the poles, the effective magnetic flux ϕ_{eff} through the rotating coil is:

$$\phi_{\text{eff}} = \phi_{mx} \cdot \sin \alpha = \phi_{mx} \cdot \sin \omega t \ [\text{V} \cdot \text{s}] \tag{4.6}$$

where

$\phi_{mx} = B \cdot A \ [\text{V} \cdot \text{s}]$ is the maximum magnetic flux the passes through the coil,
$B \ [\text{V} \cdot \text{s/m}^2]$ is the magnetic flux density, and
$A \ [\text{m}^2]$ is the coil area.

The instantaneous voltage across the coil terminals, with reference to the external circuit, can be calculated using Faraday's Law (equation (4.4)):

$$e = N \frac{d\phi_{\text{eff}}}{dt} = N \cdot \phi_{mx} \cdot \omega \cdot \cos \omega t = E_{mx} \cos \omega t \ [\text{V}] \tag{4.7}$$

where

$E_{mx} = N \cdot \phi_{mx} \cdot \omega \ [\text{V}]$ is the peak value of the induced voltage in the coil, and
N is the number of active turns in the coil.

As suggested above, the coil terminals are connected to the two detached segments of the commutator. Both, the coil and the commutator rotate together. For that reason, although the induced voltage in the armature coil reverses direction in time (equation (4.7)), the external circuit is presented with a unidirectional (rectified) voltage. That rectified voltage is shown in figure 4.3.

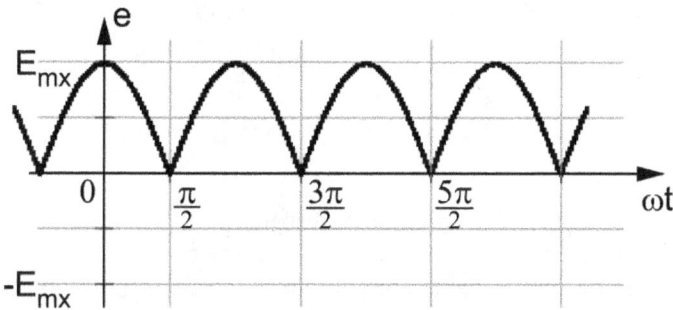

Figure 4.3. The instantaneous voltage seen by the external circuit.

The average, DC, value of the voltage wave (figure 4.3) seen by the external circuit is:

$$E_{avg} = \frac{2}{\pi} E_{mx} = \frac{2}{\pi} N \cdot \phi_{mx} \cdot \omega \ [\text{V}]$$

In practice, the machine can have more than one pole-pair. Also, the armature has multiple conductors (coils) that are evenly distributed in slots around its lateral surface. Those coils are connected in series and in parallel to multiple segments of the mechanical commutator. That brings about a momentous reduction in the ripple of the voltage waveshape seen by the external circuit. As a result:

- The *induced average voltage* E (also called the counter electromotive force, EMF) becomes:

$$E = K \cdot \phi_{mx} \cdot \omega \ [\text{V}] \tag{4.8}$$

where:

$$K = \frac{Z \cdot p}{2 \cdot \pi \cdot a} \tag{4.9}$$

and

Z is the total active conductors around the armature,
p is the number of poles, and
a is the number of parallel branches (paths) in the armature.

Notes:

The average voltage E (equation (4.8)) relates directly to the speed ω.
The product $K \cdot \phi_{mx}$ is called the EMF coefficient.
The maximum flux ϕ_{mx} relates directly to the magnetic flux density B (equation (4.1)), which relates directly to the permeability coefficient μ (equation (4.3)). Using ferromagnetic materials, that permeability value varies non-linearly when the magnetic intensity H varies.

- *The electromagnetic (EM) torque* T_{em} required to drive the armature in a generating mode, or developed by the armature in a motoring mode, can be derived from the power balance in the armature:

$$\left. \begin{array}{c} P_{mec} = P_{em} \ [\text{W}] \\ \text{or} \\ T_{em} \cdot \omega = E \cdot I_a \ [\text{W}] \end{array} \right\} \tag{4.10}$$

where
T_{em} [Nm] is the electromagnetic torque,
ω [s^{-1}] is the angular speed of the armature,
E [V] is the induced average voltage in the armature (equation (4.8)), and
I_a [A] is the average current in the armature coils.

Substituting the induced voltage E (equation (4.8)) in equation (4.10), the EM torque becomes:

$$T_{em} = \frac{E \cdot I_a}{\omega} = \frac{K \cdot \phi_{mx} \cdot \omega \cdot I_a}{\omega} = K \cdot \phi_{mx} \cdot I_a \ [\text{Nm}] \tag{4.11}$$

where I_a [A] is the average current (DC) in the armature.

4.3 Performance evaluation

As suggested above, the versatility of the speed–torque characteristics of DC motors in braking methods and in speed control is the criterion for most electromechanical drives.

4.3.1 The equivalent circuit diagram

The equivalent circuit diagram of a DC motor at given average currents in the armature and the excitation circuits is [1–3].

The excitation circuit (figure 4.4) is fed by an average current I_f [A] and generates the main magnetic field in the airgap of the machine. The ohmic resistance R_f [Ω] presents the resistance of that circuit.

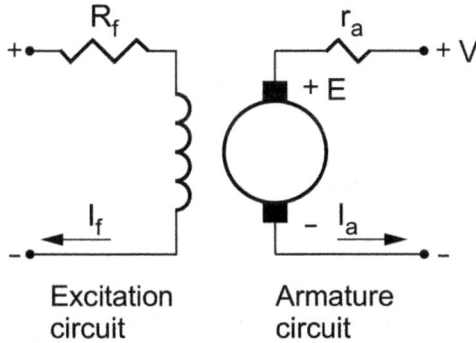

Figure 4.4. The equivalent circuit of a DC machine at given average currents.

The armature circuit (figure 4.4) generates the induced voltage E [V] and includes its internal resistance r_a [Ω]. The direction of the average current I_a indicates a motoring mode of operation. In this mode, the voltage source V supplies the power to the machine while the motor shaft produces the work (torque). In a generating mode, the average current would flow in the opposite direction. In that mode (generation), the shaft is driven by a prime mover that supplies the required torque to feed a load connected to its terminals. Applying Kirchhoff voltage law, KVL, to the armature circuit:

$$V = E \pm I_a \cdot r_a \text{ [V]} \tag{4.12}$$

where the (+) sign indicates motoring and the (−) sign indicates generating mode of operation.

4.3.2 Power and efficiency

The efficiency η of the motor is given by:

$$\eta = \frac{P_{shaft}}{P_{in}} = \frac{P_{in} - \Delta P}{P_{in}} \tag{4.13}$$

where
  P_{shaft} [W] is the output power or shaft power,
  P_{in} [W] is the input power to the motor, and
  ΔP [W] is the power losses in the motor.

The power losses are:

$$\Delta P = P_{in}(1 - \eta) = I_a^2 \cdot r_a + I_f^2 \cdot R_f + P_C + P_B + P_{mec} \ [\text{W}] \qquad (4.14)$$

where
  $I_a^2 \cdot r_a$ [W] is the copper losses in the armature,
  $I_f^2 \cdot R_f$ [W] is the copper losses in the excitation coil,
  P_C [W] is the core losses in the armature,
  P_B [W] is the power losses in the brushes, and
  P_{mec} [W] is the mechanical losses (viscous and bearing friction).

4.3.3 The armature resistance

The armature resistance might not be available in the datasheet of the machine [4–8]. In that case, its value can be estimated from the motor nameplate using two practical approximations:

(a) The excitation current is smaller than 5% of the machine nameplate current. For performance evaluation, the armature current at rated load can be considered equal to the nameplate current.
(b) The copper losses in the armature at nominal (nameplate) power are equal to half the total losses of the machine:

$$\left. \begin{array}{c} (a) \ I_{an} \cong I_n \\ \text{and} \\ (b) \ I_{an}^2 \cdot r_a \cong \dfrac{1}{2} I_n \cdot V_n(1 - \eta_n) \end{array} \right\} \ r_a \cong \frac{V_n}{2 \cdot I_n}(1 - \eta_n) \ [\Omega] \qquad (4.15)$$

where
  I_{an} [A] is the armature current at rated load,
  r_a [Ω] is the armature resistance,
  V_n [V] is the nameplate voltage,
  I_n [A] is the nameplate current, and
  η_n is the rated (nameplate) efficiency of the motor.

4.3.4 The speed–torque characteristics

The speed–torque $\omega = f(T)$ characteristic of a DC motor can be derived by substituting the induced average voltage (equation (4.8)) into equation (4.12):

$$V = K \cdot \phi_{mx} \cdot \omega + I_a \cdot r_a \ [\text{V}] \qquad (4.16)$$

and the speed as a function of the current, $\omega = f(I_a)$, is:

$$\omega = \frac{V}{K \cdot \phi_{mx}} - \frac{r_a}{K \cdot \phi_{mx}} I_a \ [\text{s}^{-1}] \tag{4.17}$$

Replacing the armature current I_a in equation (4.17) with the EM torque T_{em} (equation (4.11)), the motor speed–torque characteristic, $\omega = f(T_{em})$, becomes:

$$\left.\begin{array}{l} \omega = \dfrac{V}{K \cdot \phi_{mx}} - \dfrac{r_a}{(K \cdot \phi_{mx})^2} T_{em} = \omega_0 - \Delta\omega \ [\text{s}^{-1}] \\[6pt] \text{where} \\[6pt] \omega_0 = \dfrac{V}{K \cdot \phi_{mx}} \ [\text{s}^{-1}] \text{ is the no-load speed, and} \\[6pt] \Delta\omega = \dfrac{r_a}{(K \cdot \phi_{mx})^2} T_{em} \ [\text{s}^{-1}] \text{ is the speed drop} \end{array}\right\} \tag{4.18}$$

4.3.5 Four quadrants of operation

In a steady state condition, the resistive load–torque referred to the motor shaft is to be counter balanced by the driving motor. There are two types of load–torques: active and passive. Active torques such as gravitational forces in hoists and cranes can act in the direction of motion (loaded hoist down) or against the motion (loaded hoist up). Passive torques such as fans and friction forces are always opposed to the motion [4–8].

In view of the fact that both active and passive torques can be present in a drive system, the motor may operate in different regimes; that is, not only as a motor but, under specific conditions, also as a generator and as a brake. Moreover, the motor may be required to run in both directions. Therefore, in sketching the speed–torque characteristic of the load or the motor, it may be necessary to use all four quadrants of the speed–torque plane (figure 4.5).

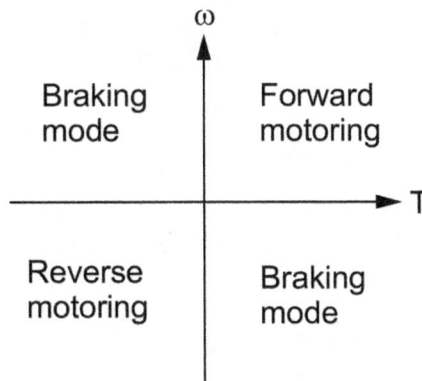

Figure 4.5. Four quadrants of operation.

- *First quadrant* (figure 4.5)—the electric motor rotates in a forward (positive) direction and develops positive torque to overcome a resistive torque required by the mechanism.
- *Second quadrant*—relating to the positive direction of speed, the motor torque acts against (braking mode) the direction of motion and the motor absorbs mechanical power converting it to electrical power. The electrical energy is fed back into the voltage source, and the braking mode is called regenerative braking mode.
- *Third quadrant*—the motor runs in the reverse (negative) direction and develops negative torque to overcome a resistive torque required by the mechanism.
- *Fourth quadrant*—relating to the positive direction of speed, the motor torque acts against (braking mode) the direction of motion and the motor absorbs mechanical power and converting it to electrical power. In this case, the electrical energy is dissipated in the resistance of the windings or in an external circuit.

4.4 Braking mode of operation

In electromechanical drives, it is frequently required to stop the working machine quickly at a precise position or at a specific angle. Moreover, some drives require an abrupt braking followed by an immediate reverse operation.

Using the electric motor in applying a braking torque to the drive has several advantages over mechanical braking: less maintenance is required; no special means are needed for cooling; the operation is smooth and chattering is avoided; the control is easier; and, the efficiency is higher.

The electric motor offers three braking methods:

(a) Regenerative braking mode—when the motor operates as a generator and electric power is returned to the energy source.

(b) Dynamic braking mode—when the motor operates as a generator but the electric power is dissipated in the motor winding or in an external resistor.

(c) Plugging braking mode or 'countercurrent braking'—when the motor is operated as a generator connected in series with the utility source and the combined electric power is wasted in the motor winding or in an external resistor.

4.4.1 Regenerative braking mode

Consider a shunt-connected DC motor where the armature and the field windings are connected in parallel. Regenerative braking mode happens when the motor shaft is driven by the working machine at a speed higher than its no-load speed. That is a braking mode at constant speed, such as in a loaded hoist down or a loaded elevator down or a downhill acceleration of an electric car.

When the shaft speed is higher than the no-load speed, $\omega > \omega_0$ (second quadrant in figure 4.6(b)), the induced average voltage, $E = K \cdot \phi_{mx} \cdot \omega$, in the armature becomes larger than the source voltage V (figure 4.6(a)). That larger induced voltage reverses the current direction in the armature, and the machine operates as a generator returning electric power back to the DC source.

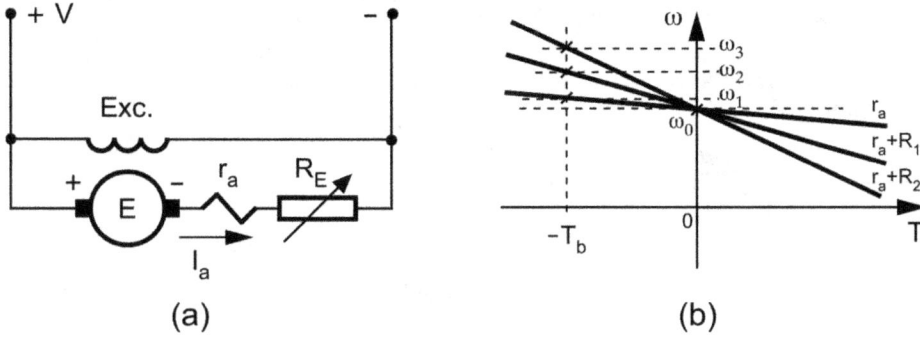

Figure 4.6. Regenerative braking of a DC motor. (a) Circuit diagram and (b) speed versus torque curves.

Applying KVL in the armature circuit suggests:

$$V + E + I_a(r_a + R_E) = 0 \quad \Longrightarrow \quad I_a = \frac{V - E}{r_a + R_E} \ [\text{A}] \tag{4.19}$$

Note that in this case: $E > V$, which causes the armature current to reverse direction.

The operation moves to the second quadrant where the EM torque (equation (4.11)) is negative. This means that the motor opposes (braking mode) the working machine from accelerating by gravitational forces. The speed–torque characteristic (equation (4.18)) becomes:

$$\omega = \frac{V}{K \cdot \phi_{mx}} - \frac{r_a + R_E}{(K \cdot \phi_{mx})^2} T_b \ [\text{s}^{-1}] \tag{4.20}$$

When the external resistors R_E increases ($R_2 > R_1 > 0$, in figure 4.6(b)), the braking speed at a given torque increases too ($\omega_3 > \omega_2 > \omega_1$).

4.4.2 Dynamic braking mode

Dynamic braking is performed by switching the armature connection from the DC source to an external resistor, while the field winding remains connected to the source (figure 4.7). That is a braking mode that brings the drive to full stop such as with cranes and propulsion systems.

Assuming forward motoring where the switch is at its upper position (figure 4.7 (a)), where the EM torque is T_1 at a speed of ω_1 (figure 4.7(b)). When the switch moves to its lower position, the armature is abruptly disconnected from the voltage source. The machine is now operating as a generator where the induced voltage E feeds both resistances, the internal r_a and the external R_E. The armature current reverses direction, and the negative EM torque opposes (braking mode) the referred load–torque.

4-11

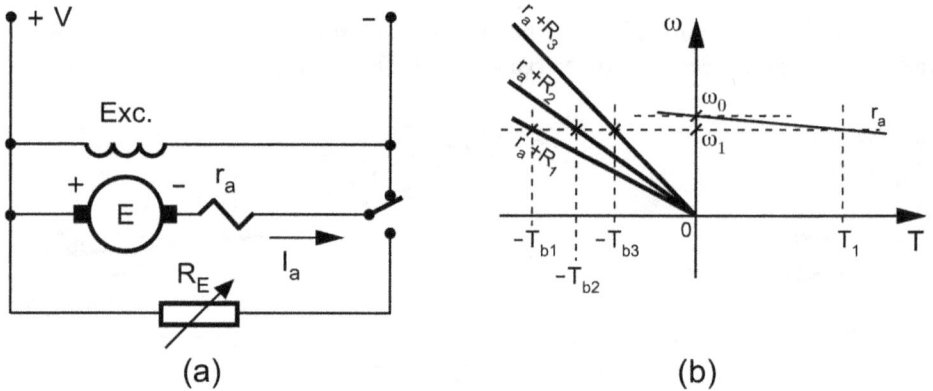

Figure 4.7. Dynamic braking of a DC motor. (a) Circuit diagram and (b) speed versus torque curves.

Applying KVL in the armature circuit suggests (switch in lower position):

$$-E - I_a(r_a + R_E) = 0 \implies I_a = -\frac{E}{r_a + R_E} \; [\text{A}] \tag{4.21}$$

The operation moves to the second quadrant where the EM torque (equation (4.11)) is negative. The motor develops (generates) a dynamic braking torque and brings the working machine to a complete stop along the speed–torque characteristic (equation (4.18)):

$$\left. \begin{aligned} \omega &= \frac{V}{K \cdot \phi_{mx}} - \frac{r_a + R_E}{(K \cdot \phi_{mx})^2} T_b \\ &\text{Since the source voltage is now: } V = 0 \\ \omega &= -\frac{r_a + R_E}{(K \cdot \phi_{mx})^2} T_b \; [\text{s}^{-1}] \end{aligned} \right\} \tag{4.22}$$

At a given speed, as the external resistor R_E increases ($R_3 > R_2 > R_1$, in figure 4.7(b)) the starting braking torque decreases ($T_{b3} < T_{b2} < T_{b1}$).

4.4.3 Plugging braking mode

Plugging braking mode is accomplished in two ways: (1) by inserting a relatively large external resistor in the armature circuit to maintain constant braking torque. That is a braking mode at constant speed, such as in lowering the load of a hoist mechanism while the motor is still connected so as to raise the load; and (2) by a reversal of the connection of the armature to the supply voltage, but not of the motor excitation, so as to bring about a quick stop.

Case 1: Inserting a relatively large external resistor in the armature circuit. The two switches are in the upper position (figure 4.8(a)). When resistor $R_E = R_1$ is inserted in the armature circuit, the EM torque T_{b1} developed by the motor becomes smaller than the referred load–torque at the motor shaft (fourth

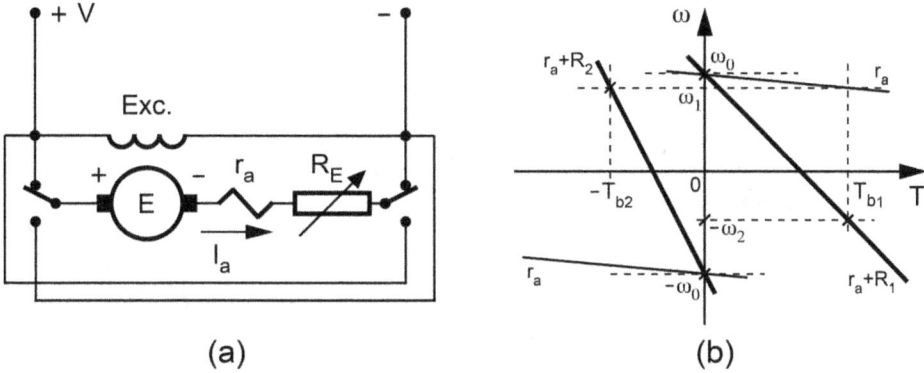

Figure 4.8. Plugging braking of a DC motor. (a) Circuit diagram and (b) speed versus torque curves.

quadrant operation), and lowering the load, as an example, is done at a constant speed, $-\omega_2$ (figure 4.8(b)).

Case 2: Reversal of the armature connection to the supply voltage. The two switches operate concurrently and move to the lower position (figure 4.8(a)). Inserting the resistor $R_E = R_2$ in the armature circuit prevents a current overloading (third quadrant operation). That switching causes an aggressive braking torque $-T_{b2}$ that brings the drive to a complete stop (figure 4.8(b)). At that point (zero speed), the motor must be disconnected from the source.

In both cases, the counter EMF, E, reverses direction and aliens with the source voltage. Applying KVL in the armature circuit suggests:

$$-V - (-E) + I_a(r_a + R_E) = 0 \implies I_a = \frac{V + E}{r_a + R_E} \ [\text{A}] \tag{4.23}$$

The speed versus torque characteristics are (figure 4.8(b)):

$$\left. \begin{array}{l} \text{In the fourth quadrant: } \omega = \dfrac{V}{K \cdot \phi_{mx}} - \dfrac{r_a + R_E}{(K \cdot \phi_{mx})^2} T_b \ [\text{s}^{-1}] \\[3mm] \text{In the second quadrant: } \omega = \dfrac{-V}{K \cdot \phi_{mx}} - \dfrac{r_a + R_E}{(K \cdot \phi_{mx})^2} T_b \ [\text{s}^{-1}] \end{array} \right\} \tag{4.24}$$

Varying the external resistor R_E, the braking torque T_b and the speed ω would also vary.

Example 4.2 A shunt-connected DC motor has the following nameplate values: 14.6 kW. 300 V, 76 A, 886 rpm, and the resistance of the excitation circuit is 75 Ω.

Assume constant magnetic flux in the airgap of the machine and constant iron and friction losses. Also, neglect the brush losses, and address the following questions:

(a) The motor operates at the regenerative braking mode (second quadrant; figure 4.6), and returns energy to the supply circuit. Calculate the speed of the shaft ω_1 when the armature current is $I_a = -60$ A, without any external resistor connected in the armature circuit (natural characteristic of the motor).

(b) Dynamic braking mode is applied to stop the motor from a 500 rpm shaft-speed (second quadrant; figure 4.7). Calculate the external resistance inserted in the armature circuit to limit the current to its rated value. Also, calculate the braking torque at the motor shaft at that speed.

(c) The motor operates in a plugging braking mode (fourth quadrant; figure 4.8) at a speed of 600 rpm, and armature current of 50 A. Calculate the external resistance R_1 inserted in the armature circuit; the EM torque T_{b1}; the shaft power; the input power to the motor; and the power dissipated in the armature circuit.

Solution

Motor parameters:

The rated motor efficiency is:

$$\eta_n = \frac{P_{shaft}}{P_{in}} = \frac{14.6 \cdot 10^3}{300 \cdot 76} = 0.64$$

The armature resistance (equation (4.15)):

$$r_a \cong \frac{V_n}{2 \cdot I_n}(1 - \eta_n) = \frac{300}{2 \cdot 76}(1 - 0.64) = 0.71 \ [\Omega]$$

The current in the excitation circuit is:

$$I_f = \frac{V}{R_f} = \frac{300}{75} = 4 \ [A]$$

The rated armature current is:

$$I_{an} = I_n - I_f = 76 - 4 = 72 \ [A]$$

The rated radial speed is:

$$\omega_n = \frac{2\pi \cdot n_n}{60} = \frac{2\pi \cdot 886}{60} = 92.8 \ [s^{-1}]$$

Using equation (4.17), the EMF coefficient is:

$$\omega_n = \frac{V_n}{K \cdot \phi_{mx}} - \frac{r_a}{K \cdot \phi_{mx}}I_{an} \implies K \cdot \phi_{mx} = \frac{300 - 72 \cdot 0.71}{92.8} = 2.68$$

The no-load radial speed is (equation (4.18)):

$$\omega_0 = \frac{V_n}{K \cdot \phi_{mx}} = \frac{300}{2.68} = 111.9 \ [s^{-1}]$$

The rated EM torque is (equation (4.11)):

$$T_{em, \ n} = K \cdot \phi_{mx} \cdot I_{an} = 2.68 \cdot 72 = 193.1 \ [Nm]$$

The rated shaft torque is (equation (1.7)):

$$T_n = \frac{P_n}{\omega_n} = \frac{14.6 \cdot 10^3}{92.8} = 157.3 \ [Nm]$$

The torque losses are:

$$\Delta T = T_{em,\ n} - T_n = 193.1 - 157.3 = 35.8 \ [\text{Nm}]$$

(a) *Regenerative braking mode* (second quadrant; figure 4.6(b))
   The EM braking torque is:

$$T_b = K \cdot \phi_{mx} \cdot I_a = 2.68 \cdot (-60) = -160.8 \ [\text{A}]$$

The shaft speed is (equation (4.20) where $R_E = 0$):

$$\omega_1 = \frac{V_n}{K \cdot \phi_{mx}} - \frac{r_a}{(K \cdot \phi_{mx})^2} T_b = \frac{300}{2.68} - \frac{0.71}{2.68^2}(-160.8) = 127.8 \ [\text{s}^{-1}]$$

Assuming constant torque losses, the torque at the motor-shaft is:

$$T_{shaft} = T_b + \Delta T = 160.8 + 35.8 = 196.6 \ [\text{Nm}]$$

During braking mode, the load drives the motor. Therefore, the motor losses are covered by the load.

(b) *Dynamic braking mode* (second quadrant; figure 4.7(b)).
   The starting radial speed is:

$$\omega_1 = \frac{2\pi \cdot 500}{60} = 52.4 \ [\text{s}^{-1}]$$

The external resistance inserted in the armature circuit is (equation (4.22)):

$$\omega = -\frac{r_a + R_E}{(K \cdot \phi_{mx})^2} T_b \implies . \quad R_E = -\frac{52.4 \cdot 2.68^2}{-193.1} - 0.71 = 1.24 \ [\Omega]$$

Assuming constant torque losses, the torque at the motor-shaft is:

$$T_{shaft} = T_b + \Delta T = 193.1 + 35.8 = 228.9 \ [\text{Nm}]$$

(c) *Plugging braking mode* (fourth quadrant in figure 4.8(b))
   The radial speed during braking conditions is:

$$\omega_2 = \frac{2\pi \cdot 600}{60} = 62.8 \ [\text{s}^{-1}]$$

The EM braking torque is:

$$T_{b1} = K \cdot \phi_{mx} \cdot I_a = 2.68 \cdot 50 = 134 \ [\text{A}]$$

The induced voltage in the armature is (equation (4.8)):

$$E = =K \cdot \phi_{mx} \cdot \omega_2 = 2.68 \cdot 62.8 = 168.3 \ [\text{V}]$$

The external resistor inserted in the armature circuit is (equation (4.23)) is:

$$I_a = \frac{V + E}{r_a + R_1} \implies R_1 = \frac{300 + 168.3}{50} - 0.71 = 8.66 \ [\Omega]$$

Assuming constant torque losses, the torque at the motor-shaft is:

$$T_{shaft} = T_{b1} + \Delta T = 134 + 35.8 = 169.8 \ [\text{Nm}]$$

The power at the motor-shaft is (supplied by the mechanical load):

$$P_{shaft} = T_{shaft} \cdot \omega_2 = 169.8 \cdot 62.8 = 10.66 \ [\text{kW}]$$

The input power to the motor during brake is:

$$P_{in} = V_n(I_a + I_f) = 300(50 + 4) = 16.2 \ [\text{kW}]$$

The copper losses in the armature are:

$$P_{cu} = I_a^2 \cdot r_a = 50^2 \cdot 0.71 = 1.78 \ [\text{kW}]$$

The power losses in the external resistor are:

$$P_{ext} = I_a^2 \cdot R_1 = 50^2 \cdot 8.66 = 21.65 \ [\text{kW}]$$

The iron and friction losses are:

$$P_{\text{iron+friction}} = (P_{in} + P_{shaft}) - (P_{cu} + P_{ext}) =$$
$$= (16.2 + 10.66) - (1.78 + 21.65) = 3.43 \ [\text{kW}]$$

Check: The shaft power can also be calculated as

$$\boldsymbol{P_{shaft} = P_{in} - \Delta P} = 16.2 - (21.65 + 1.78 + 3.43) = -10.66 \ [\text{kW}]$$

The same shaft power as was calculated above. The negative sign suggests that the shaft-power is supplied to the motor.

4.5 Aspects of speed control

Industrial state-of-the art electromechanical drives employ machines that require different and adjustable speeds, such as electric traction systems, machine tools, textile industries, and domestic water systems.

Main features:

The fundamental features by which various methods of electric speed control can be characterized are:

(a) *Range of speed control K_R is defined as:*

$$K_R = \frac{\omega_{mx}}{\omega_{min}} \tag{4.25}$$

where

$\omega_{mx} \ [\text{s}^{-1}]$ is the maximum required speed of the drive and
$\omega_{min} \ [\text{s}^{-1}]$ is the minimum required speed.
Continuity (smoothness) of speed control K_C is defined:

$$K_C = \frac{\omega_n}{\omega_{n-1}} \tag{4.26}$$

where

1. ω_n [s^{-1}] is the steady speed of the nth step of speed control and
2. ω_{n-1} [s^{-1}] is the steady speed of the $(n-1)$th step.
3. *Economic rational* depends first on the capital cost of the drive system and second on its viable efficiency η_V at different speeds:

$$\eta_V = \sum_{j=1}^{n} \frac{P_j \cdot t_j}{(P_j + \Delta P_j) \cdot t_j} \qquad (4.27)$$

where

j is a time interval at a given speed,

P_j [W] is the power at the motor shaft at interval j,

ΔP_j [W] is the power loss in the motor and speed control unit at interval j, and

t [s] is the time.

4. *Stability of operation* at a given speed is defined as the change in speed caused by a change in the load–torque. For a given mechanism, the harder the motor-speed characteristic is (see section 1.3.1), the greater the stability.
5. *Direction of the speed control* is defined as the direction in which the speed of the motor is changed from a base speed and depends of the method of control.

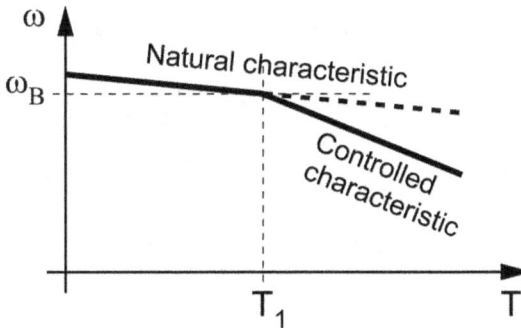

Figure 4.9. Designation of base speed.

In figure 4.9, the base speed ω_B is a designed speed that the motor develops at rated voltage for a predetermined value of load–torque T_1.

6. *The permissible load* at the motor-shaft during speed control is determined by the rated current of the motor. Each method of speed control corresponds to specific characteristic of the load.

- One example is a crane machine during hoisting (figure 4.10(a)) that requires speed control at constant torque; that is, the shaft torque T_{shaft}^{D} is given by:

$$T_{shaft}^{D} = F \cdot r \text{ [Nm]}$$

and is constant at all speeds. The shaft power P_{shaft}^D would be:

$$P_{shaft}^D = T_{shaft}^D \cdot \omega \ [\text{W}]$$

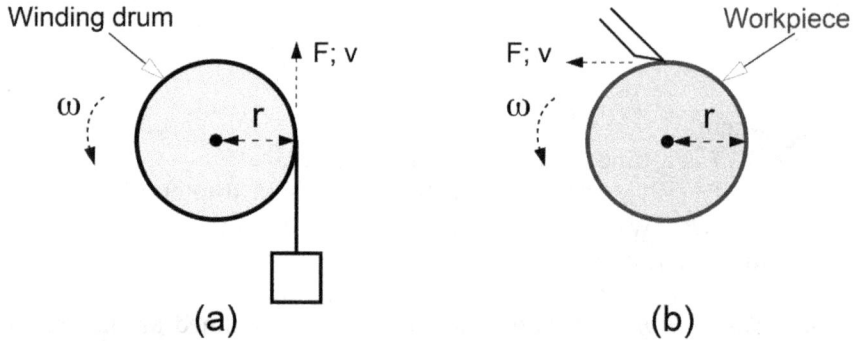

Figure 4.10. Examples of two type of loads. (a) Crane machine during hoisting, and (b) metal-cutting lathe.

- A second example is a metal-cutting lathe (figure 4.10(b)) that requires speed control at constant power; that is, the linear velocity of the workpiece surface is kept constant to sustain a constant cutting speed v and force F. The shaft power P_{shaft}^W is given by:

$$P_{shaft}^W = F \cdot v \ [\text{W}]$$

and is constant at all speeds. The shaft torque T_{shaft}^W would be:

$$T_{shaft}^W = \frac{P_{shaft}^W}{\omega} \ [\text{Nm}]$$

4.6 Speed control of DC motors

The speed–torque formula (equation (4.18)) suggests three ways of varying the speed of a DC motor [1–3]:

(a) varying an external resistance, $R_1 > R_2$, that can be inserted in the armature circuit (connected in series with the armature resistance r_a). The no-load speed ω_0 remains constant, but the speed drop $\Delta\omega$ varies (figure 4.11(a)).

(b) varying the main magnetic flux $\phi_{mx} > \varnothing_{mx1} > \varnothing_{mx2}$. Both, the no-load speed ω_0 and the speed drop $\Delta\omega$ vary (figure 4.11(b)).

(c) varying the input voltage V supplied to the armature. The no-load speed ω_0 varies, but the speed drop $\Delta\omega$ remains constant (figure 4.11(c)).

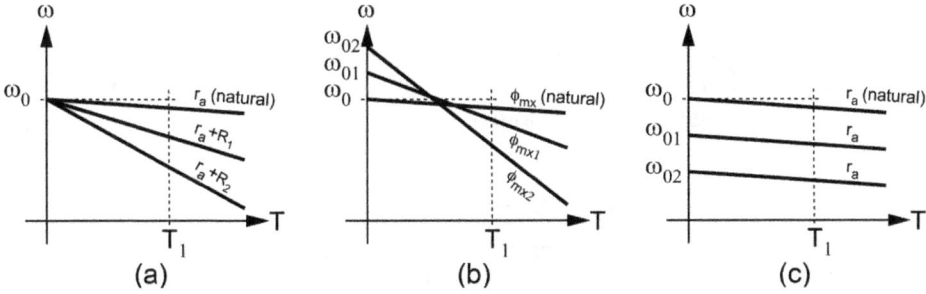

Figure 4.11. Speed control of DC motor. (a) External resistance, (b) variable magnetic flux, and (c) variable armature voltage.

The hardness of each characteristic is measured by the tangent of the curve at a given torque. When a resistance is added in the armature circuit (figure 4.11(a)), the slope of the curve increases as the resistance gets larger. When the main magnetic field is reduced (figure 4.11(b)), the slope increases drastically. When varying the voltage supplied to the armature (figure 4.11(c)), the slope remains the same for all curves.

Of the above three methods, the third one, *varying the armature voltage* (figure 4.11(c)), is the most desirable control method. At constant flux, $\phi_{mx} = $ Const., which can be realized by external excitation, only the no-load speed ω_0 varies:

$$\left. \omega_0 = \frac{V_n}{K \cdot \phi_{mx}} \; ; \; \omega_{01} = \frac{V_1}{K \cdot \phi_{mx}} \; ; \; \omega_{02} = \frac{V_2}{K \cdot \phi_{mx}} \atop \text{where } V_n \text{ is the rated voltage, and} \atop V_n > V_1 > V_2 \right\} \qquad (4.28)$$

Using external excitation, the variation in armature voltage would not affect the slope of the curves, that is $r_a/(K \cdot \phi_{mx})^2 = $ Const. in equation (4.18). As a result, the method of varying the armature voltage produces a family of linear hard speed–torque characteristics parallel to the motor natural characteristic (figure 4.11(c)).

At a given torque, the speed versus armature voltage curve, $\omega = f(V)$, would also exhibit a linear behavior (figure 4.12).

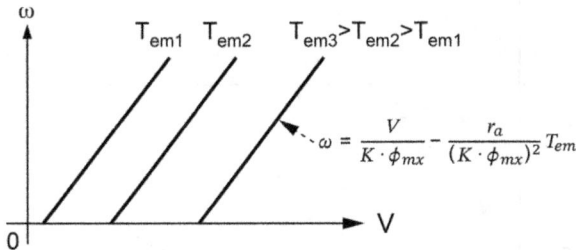

Figure 4.12. Speed–voltage characteristic of a DC motor, where $\phi_{mx} = $ const.

That speed-control method, varying the armature input voltage (figures 4.11(c) and 4.12), is the criterion for most electromechanical drives. The following section addresses a DC voltage-controlled source for that purpose.

4.6.1 AC/DC phase-controlled rectifiers

A practical way of varying the armature voltage of DC motors, in order to control their speed, is applying single- or three-phase AC/DC phase-controlled bridge rectifiers [9–13].

The most common *phase-controlled rectifiers* are the line-commutated ones. The switching devices are thyristors and switching of the current from one device to the next is achieved by the source AC voltages, which means natural commutation (as opposed to forced commutation that requires an external circuit to commutate the device).

A sample of the instantaneous output voltage (DC voltage) of a few *uncontrolled bridge rectifiers* is shown in figure 4.13.

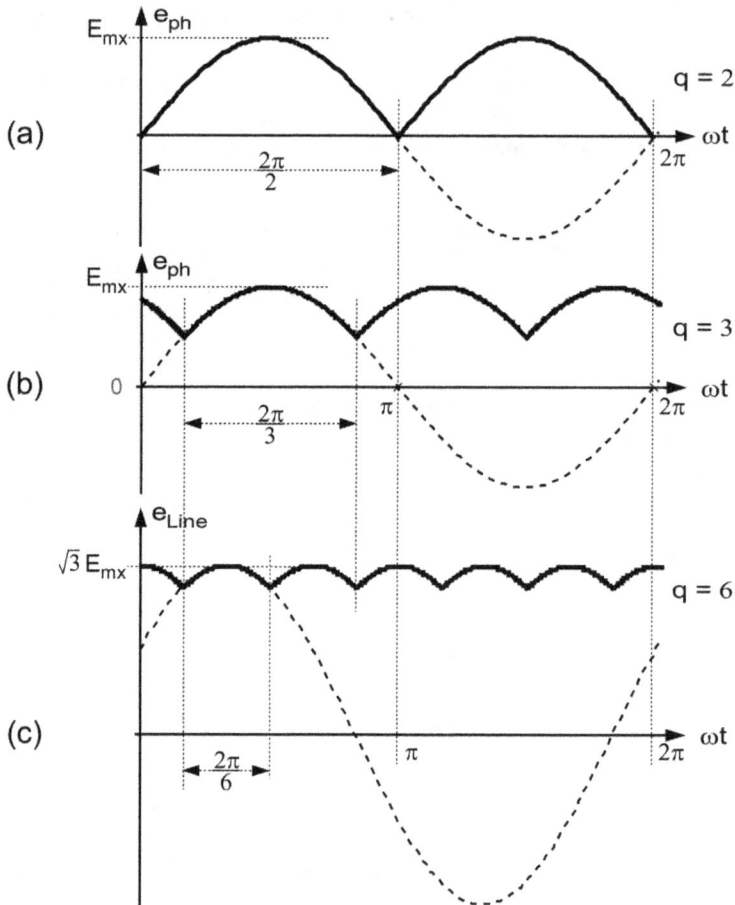

Figure 4.13. DC voltage of uncontrolled bridge rectifiers: (a) single-phase full-bridge, two pulses; (b) three-phase mid-point, three pulses; and (c) three-phase full-bridge, six pulses.

The number of DC voltage-pulses per source-cycle, 2π, depends on the configuration of the bridge. A single-phase, bridge rectifier generates two pulses per cycle, $q = 2$, where the peak voltage E_{mx} is the magnitude of the source phase-voltage (figure 4.13(a)). A three-phase mid-point connection rectifier generates three pulses, $q = 3$, where E_{mx} is also the magnitude of the source phase-voltage (figure 4.13(b)). A three-phase bridge rectifier generates six pulses, $q = 6$, where the peak voltage $\sqrt{3}\ E_{mx}$ is the magnitude of the source line-voltage (figure 4.13(c)).

Assuming that the phase voltage is given by $e_a = E_{mx} \cos \omega t$, the ideal (no losses) average voltage V_{do} of two or more voltage-pulses can be calculated as follows:

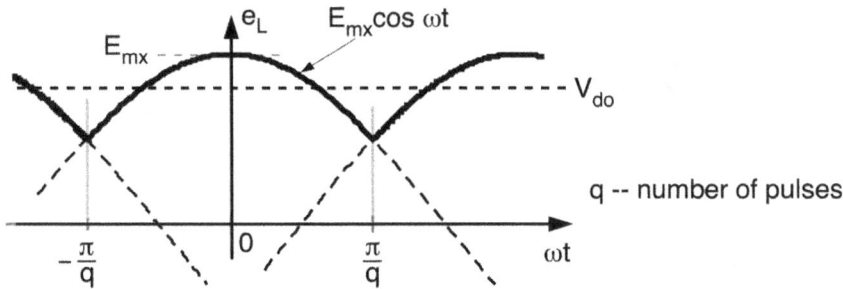

Figure 4.14. Instantaneous output voltage of an uncontrolled rectifier.

The *ideal average (DC) voltage* across the load for $q \geqslant 2$, where the pulse-cycle is $2\pi/q$ (figure 4.14), is:

$$V_{d0} = \frac{1}{\frac{\pi}{q}} \int_0^{\frac{\pi}{q}} E_{mx} \cos \omega t \cdot d(\omega t) = E_{mx} \frac{q}{\pi} \sin \frac{\pi}{q} \ \text{[V]} \tag{4.29}$$

Note that E_{mx} (equation (4.29)) is the peak phase-voltage. For a $3 - \phi$ bridge rectifier, the peak line-voltage would be $\sqrt{3}\ E_{mx}$ (figure 4.13(c)).

The schematic diagram of a 3-ϕ **phase-controlled** AC/DC mid-point connection rectifier is shown in figure 4.15(a).

The 3-ϕ secondary instantaneous voltages, e_{1-0}, e_{2-0}, and e_{3-0} are sketched in figure 4.15(b). For the sake of argument, assume that the reactance of the load is much larger than its ohmic resistance, $\omega L \gg R$, which suggests that the current I_d in the load can be considered constant (figure 4.15(a)).

Consider that thyristor T_1 conducts the full load current I_d, while thyristors T_2 and T_3 are negatively biased (figure 4.15(b)). Therefore, the voltage across the load is e_{1-0}. At the intersection with the second phase-voltage, that is at $\omega t = 5\pi/6$, the voltage e_{2-0} becomes larger than e_{1-0}. Although thyristor T_2 now sees positive bias, it cannot commence conducting without a triggering pulse. Therefore, thyristor T_1 continues to carry the full load current.

After a phase delay of α radians (figure 4.15(b)), at $\omega t = 5\pi/6 + \alpha$, thyristor T_2 is triggered on forcing a negative bias across T_1, because $e_{2-0} > e_{1-0}$ (natural commutation). Thyristor T_2 would now carry the full load current, and the voltage across the load would be e_{2-0}. The same commutation process is applied at each voltage intersection; for

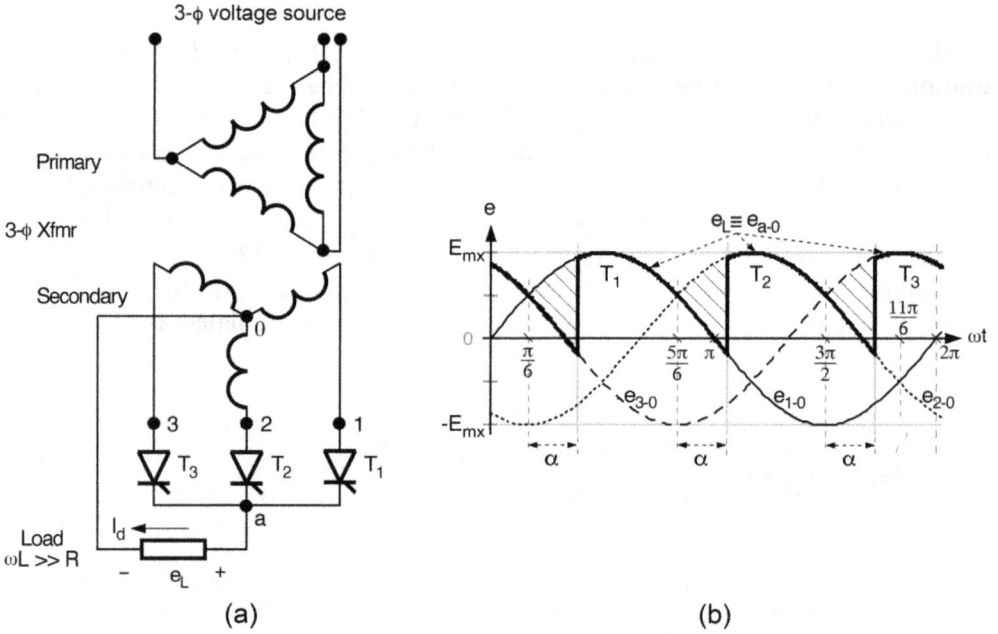

Figure 4.15. 3-ϕ mid-point connection controlled rectifier. (a) Circuit diagram and (b) instantaneous voltages.

instance, at $\omega t = 3\pi/2 + \alpha$, thyristor T_3 would pick up the load current from T_2, and at $\omega t = \pi/6 + \alpha$, thyristor T_1 would pick up the load current from T_3.

- α is defined as the delayed firing angle. Its reference point, $\alpha = 0$, is at the intersection between two phase-voltages. Note that $\alpha = 0$ presents an equivalent case to a simple diode rectifier operation.

  The practical range for the delayed firing angle depends on the bias applied across the thyristor. For instance, thyristor T_2 sees positive bias, $e_{2-0} > e_{1-0}$, during the interval $5\pi/6 \leqslant \omega t < 11\pi/6$ (figure 4.15(b)), which is half a source-cycle. Therefore, the practical range for α is:

$$0 \leqslant \alpha < \pi \tag{4.30}$$

The average voltage (DC voltage) caused by the phase-delay can be obtained from the instantaneous behavior of the load voltage (figure 4.16).

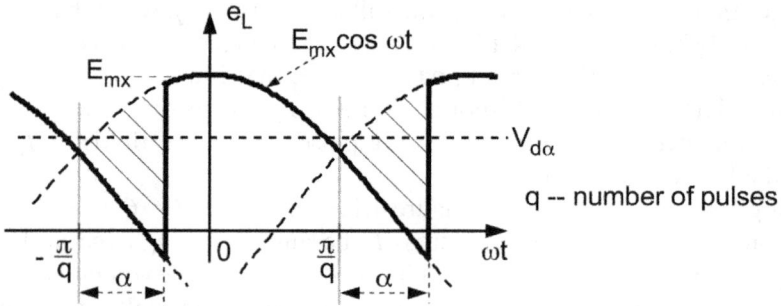

Figure 4.16. Controlled rectifier: instantaneous voltage across the load for $q \geqslant 2$.

Neglecting losses in the transformer, the average voltage $V_{d\alpha}$ across the load for the phase-controlled rectifier, where $q \geqslant 2$ is:

$$
\left.
\begin{aligned}
V_{d\alpha} &= \frac{E_{mx}}{\frac{2\pi}{q}} \int_{-\frac{\pi}{q}+\alpha}^{\frac{\pi}{q}+\alpha} \cos \omega t \cdot d(\omega t) \\
&= \frac{E_{mx}}{\frac{2\pi}{q}} \left\{ \sin\frac{\pi}{q}\cos\alpha + \sin\alpha\cos\frac{\pi}{q} - \left[\sin\left(-\frac{\pi}{q}\right)\cos\alpha + \sin\alpha\cos\left(-\frac{\pi}{q}\right) \right] \right\} \\
&\text{Finally: } V_{d\alpha} = V_{d0} \cdot \cos\alpha \ [V]
\end{aligned}
\right\} \quad (4.31)
$$

where

   $V_{d\alpha}$ is the average voltage caused by the phase-delay, and
   V_{d0} is the ideal average voltage (equation (4.29)).

Note that during the interval $\pi/2 < \alpha < \pi$, the average output voltage reverses direction, that is: $V_{d\alpha} < 0$ (equation (4.31)). It means that for a constant load current, average power is supplied back to the AC source.

The discussion above assumed ideal components. In practice, the transformer leakage inductance and the Thevenin's equivalent reactance of the network feeding the transformer transferred to the secondary winding generate reactive voltage drop that cannot be ignored. The combined reactance is called the commutating reactance x_c.

In high power applications, the most common rectifier is the six-pulse phase-controlled rectifier (figure 4.17). The schematic diagram includes the reactance x_c.

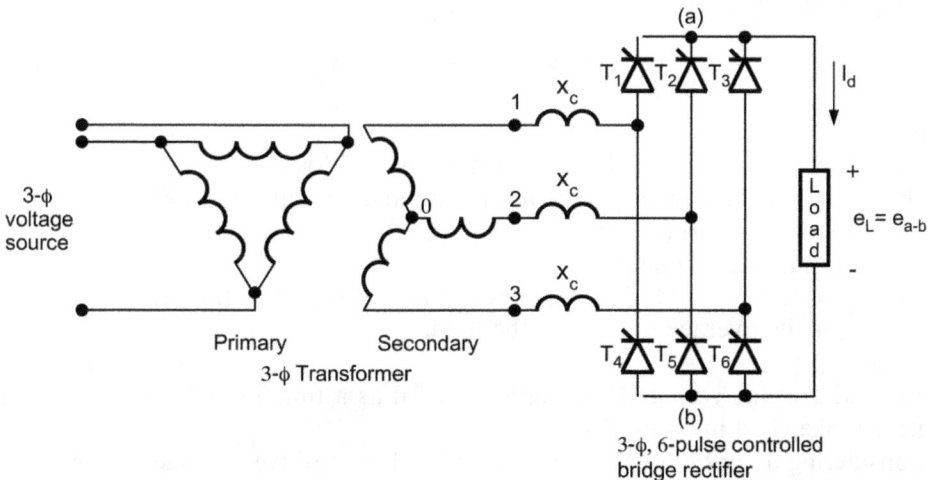

Figure 4.17. Six-pulse phase-controlled rectifier fed by a $3 - \phi \, \Delta/Y$ transformer.

The Δ/Y transformer (figure 4.17) feeds the thyristor bridge through the commutating reactance x_c connected to its secondary windings. At constant load current I_d, the instantaneous voltage e_L across the load is (figure 4.18):

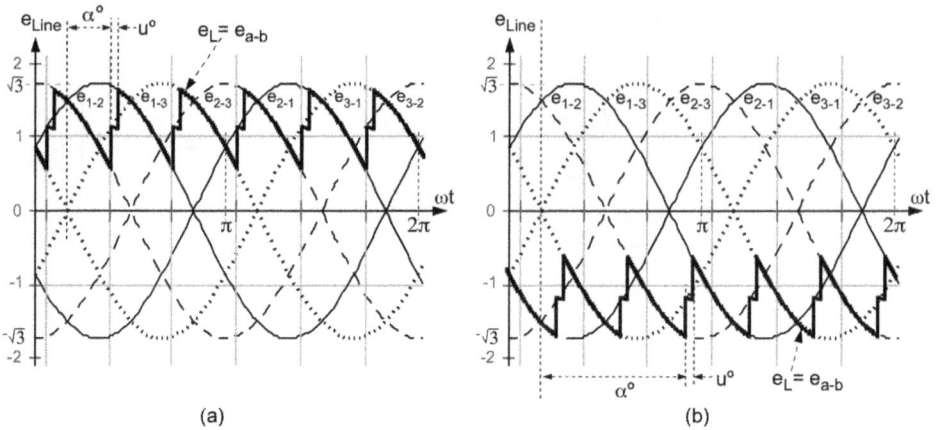

Figure 4.18. Voltage e_L across the load at continuous current and a firing angle. (a) $0° \leqslant \alpha < 90°$, and (b) $90° \leqslant \alpha < 180°$.

The reactive voltage-drop affects the waveshape of the instantaneous voltage across the load. Following a delayed firing angle α, an overlap current-event occurs between two adjacent thyristors and the interval of that event is displayed by an overlap angle $u°$. The instantaneous voltage across the load is sketched by a bold solid line in figure 4.18(a) (positive average voltage), and in figure 4.18(b) (negative average voltage).

The actual average voltage V_d between terminals a and b (across the load in figure 4.18) becomes [9–13]:

$$V_d = V_{d0} \cos \alpha - \frac{q \cdot x_c}{2\pi} I_d \ \ [\text{V}] \tag{4.32}$$

where

V_d is the actual average (DC) voltage across the load,
V_{d0} is the ideal average voltage across the load (equation (4.29)),
α is the delayed firing angle,
q is the number of pulses,
X_c [Ω] is the commutating reactance at the terminals of the rectifier, and
I_d [A] is the average current in the load.

The actual average voltage V_d (equation (4.32)) as a function of the delayed firing angle α is sketched in figure 4.19.

Considering a continuous current in the load. A positive average voltage $V_d > 0$ (equation (4.32)) indicates that the power flows from source to load. A negative average voltage $V_d < 0$, indicates that the power flows from load to source.

Note: In practice, to maintain continuous current in the load at negative average voltages, the load must have its own power source.

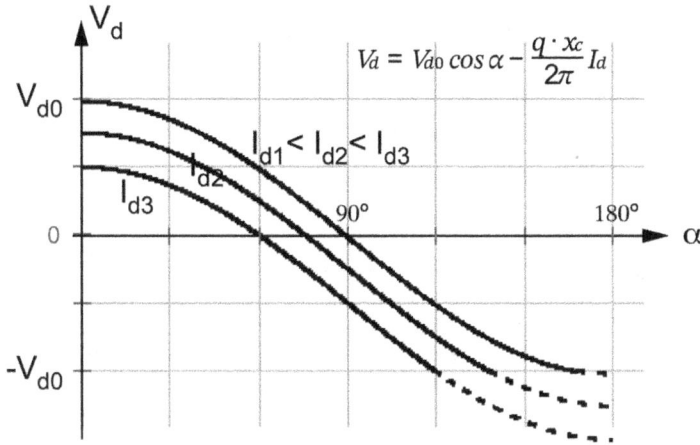

Figure 4.19. Average voltage across the load as a function of the delayed firing angle.

The equivalent diagram of a six-pulse phase-controlled rectifier operating at constant load current (figure 4.20) is:

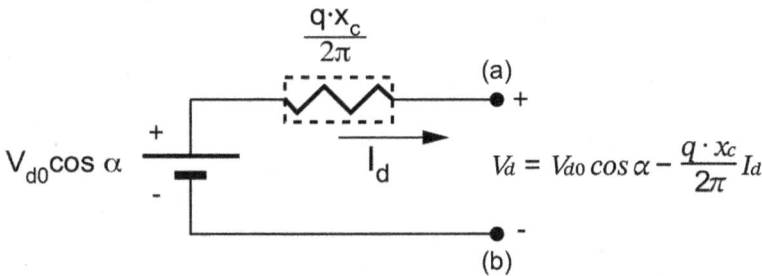

Figure 4.20. The equivalent diagram of the six-pulse phase-controlled rectifier.

Note that the component $q \cdot x_c / 2\pi$ (figure 4.20) is an equivalent ohmic resistance (not a reactance).

4.6.2 DC motor powered by a phase-controlled rectifier

Replacing the load in figure 4.17 with the armature of the DC motor would enable speed control of the motor. The excitation circuit is separately fed from another DC source, usually at the nameplate (nominal) field current. Following, the equivalent diagram of the DC motor (figure 4.4) can now be added to the equivalent diagram of the phase-controlled rectifier (figure 4.17). The merged equivalent diagram would be (figure 4.21).

Applying KVL (figure 4.21) to derive the speed of the motor vs the load (armature) current:

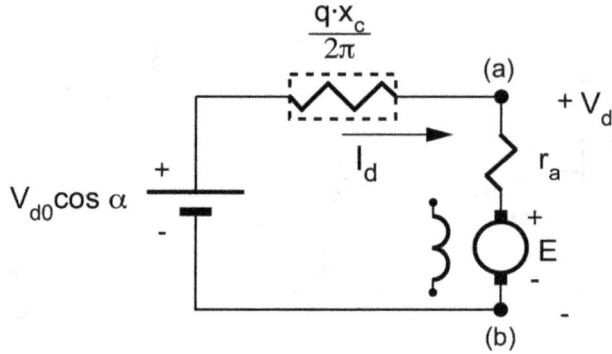

Figure 4.21. An equivalent diagram of a phase-controlled rectifier connected to a DC motor having external (separate) field excitation.

$$\left.\begin{array}{c} -V_{d0}\cos\alpha + \dfrac{q \cdot x_c}{2\pi}I_d + r_a \cdot I_d + E = 0 \\[2mm] \text{equation (4.8):} \quad E = K \cdot \phi_{mx} \cdot \omega \\[2mm] \text{Finally:} \\[2mm] \omega = \dfrac{V_{d0}}{K\phi_{mx}}\cos\alpha - \dfrac{\dfrac{q \cdot x_c}{2\pi} + r_a}{K\phi_{mx}}I_d \;\; [\text{s}^{-1}] \end{array}\right\} \qquad (4.33)$$

Replacing the armature (load) current ($I_a \equiv I_d$) in equation (4.33) with the EM torque T_{em} (equation (4.11)), the motor speed $\omega = f(T_{em})$ becomes:

$$\omega = \frac{V_{d0}}{K\phi_{mx}}\cos\alpha - \frac{\dfrac{q \cdot x_c}{2\pi} + r_a}{(K\phi_{mx})^2}T_{em} \;\; [\text{s}^{-1}] \qquad (4.34)$$

When the delayed firing angle α varies, $0 \leqslant \alpha < \pi$, the speed–torque characteristics presents a four-quadrant operation of the DC motor (figure 4.22).

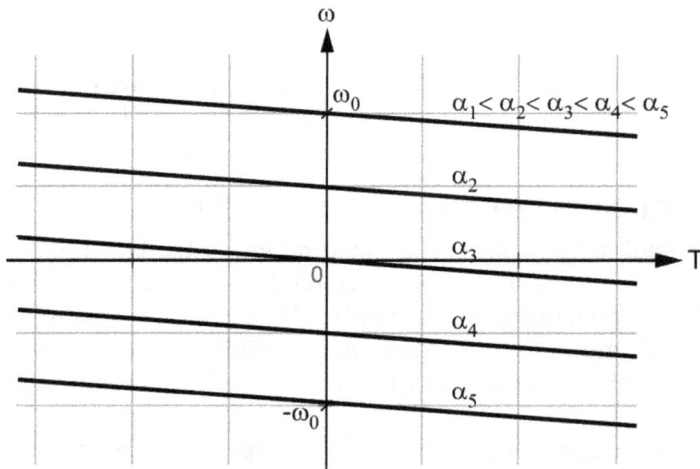

Figure 4.22. Four-quadrant operation of a DC motor: speed vs torque.

In practice, to obtain a third quadrant operation (figure 4.22), a dual converter must be applied. That is, a reverse controlled bridge rectifier has to be added to the system:

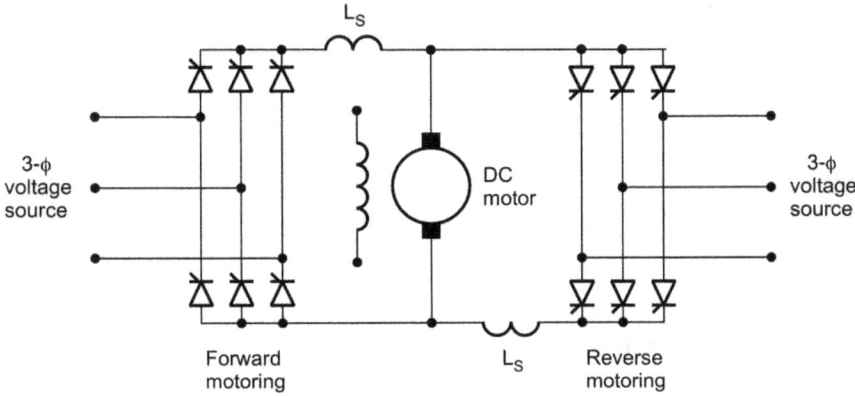

Figure 4.23. Four-quadrant operation of a DC motor: dual converter.

Operation in the first quadrant is controlled by the forward motoring converter (figure 4.23), while operation in the third quadrant is done by the reverse motoring converter. The added small inductances L_S in DC links provide a practical time-delay when switching from one converter to the other.

Example 4.3 A separately excited DC motor (the excitation field is supplied separately at constant current) is fed by an AC/DC, 3-ϕ, 6-pulse, phase-controlled rectifier (figure E4.2). The motor nameplate is: 120 V, 30 A, 3.24 kW, and 970 rpm.

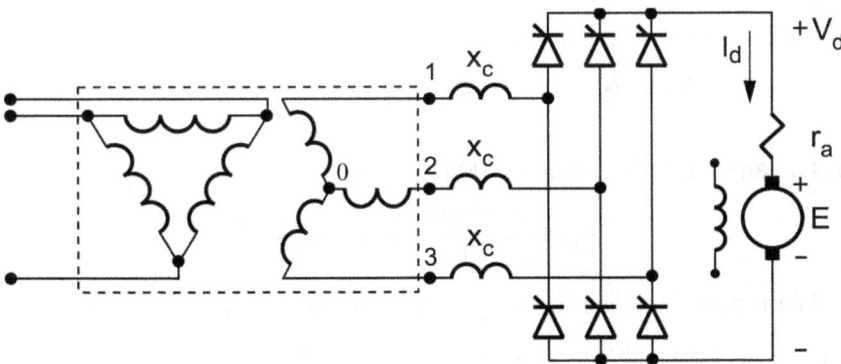

Figure E4.2. DC motor fed by a 3-ϕ, 6-pulse, phase-controlled rectifier.

4-27

The secondary phase-voltage of the 3-ϕ transformer is: $V_{1-0} = 120$ V, and the commutating reactance is: $X_c = 0.2 \, \Omega/\text{ph}$.

The controlled phase angle α is set to obtain a shaft speed of $n_1 = 500$ rpm at rated current to the motor. Neglect the excitation current and assume constant current in the armature circuit. Calculate:

(a) The induced voltage E in the armature at n_1.
(b) The average voltage across the armature at n_1.
(c) The required controlled phase angle α for n_1.

Solution

The motor parameters are:
The rated efficiency is:

$$\eta_n = \frac{P_n}{V_n \cdot I_n} = \frac{3240}{120 \cdot 30} = 0.9$$

The armature resistance is (equation (4.15)):

$$r_a \cong \frac{V_n}{2 \cdot I_n}(1 - \eta_n) = \frac{120}{2 \cdot 30}(1 - 0.9) = 0.2 \ [\Omega]$$

The rated angular speed of the motor is:

$$\omega_n = \frac{2\pi \cdot n_n}{60} = \frac{2\pi \cdot 970}{60} = 101.58 [\text{s}^{-1}]$$

The rated value of the product $K \cdot \phi_{mx}$ is (equation (4.17)):

$$\omega_n = \frac{V_n}{K \cdot \phi_{mx}} - \frac{r_a}{K \cdot \phi_{mx}} I_a \implies K \cdot \phi_{mx} \cong \frac{120 - 30 \cdot 0.2}{101.58} = 1.122 \ [\text{Vs}]$$

The no-load angular speed is:

$$\omega_0 = \frac{V_n}{K \cdot \phi_{mx}} = \frac{120}{1.122} = 106.95 \ [\text{s}^{-1}]$$

The rated EM torque is (equation (4.11)):

$$T_{em} = K \cdot \phi_{mx} \cdot I_a \cong 1.122 \cdot 30 = 33.66 \ [\text{Nm}]$$

(a) The angular speed at $n_1 = 500$ [rpm] is:

$$\omega_{500} = \frac{2\pi \cdot 500}{60} = 52.36 \ [\text{s}^{-1}]$$

The induced voltage in the armature at ω_{500} is (equation (4.8)):

$$E_{500} = K \cdot \phi_{mx} \cdot \omega_{500} = 1.122 \cdot 52.36 = 58.75 \ [\text{V}]$$

(b) The average voltage across the motor terminals at ω_{500} is:

$$V_{d,\ 500} = E_{500} + I_a \cdot r_a \cong 58.75 + 30 \cdot 0.2 = 64.75 \text{ [V]}$$

(c) The ideal average voltage of the rectifier is (equation (4.29)):

$$V_{d0} = E_{mx} \frac{q}{\pi} \sin \frac{\pi}{q} = \sqrt{2}\sqrt{3} \cdot 120 \frac{6}{\pi} \sin \frac{\pi}{6} = 280.7 \text{ [V]}$$

The required controlled phase angle is (equation (4.33)):

$$\omega_{500} = \frac{V_{d0}}{K\phi_{mx}} \cos \alpha - \frac{\dfrac{q \cdot x_c}{2\pi} + r_a}{K\phi_{mx}} I_d$$

$$52.36 = \frac{280.7}{1.122} \cos \alpha - \frac{\dfrac{6 \cdot 0.2}{2\pi} + 0.2}{1.122} 30 \implies \alpha = 75.46°$$

Check (equation (4.24)):

$$V_{d,\ 500} = V_{d0} \cos \alpha - \frac{q \cdot x_c}{2\pi} I_d = 280.7 \cdot \cos 75.46° - \frac{6 \cdot 0.2}{2\pi} 30 = 64.74 \text{ [V]}$$

Practically the same result as in (b).

4.7 Problems

1. A shunt-connected DC motor has the following nameplate data: 4 kW, 120 V, 40 A, and 955. rpm. The motor drives a hoist machine (a winding drum) via a gearbox having a gear ratio $i = 10$, and efficiency $\eta = 0.9$ (figure P4.1). The diameter of the drum is $D = 0.5$ m. The load weighs $G = 100$ kg$_f$.

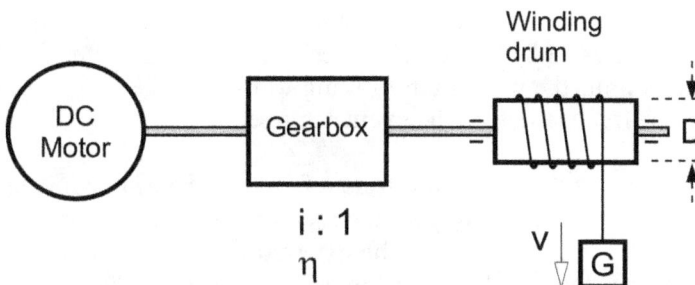

Figure P4.1. Hoist machine driven by a DC motor.

The motor is connected in a downward motion to lower the load at constant velocity. Neglect the excitation losses and calculate the downward velocity v.

2. A separately excited DC motor has the following specifications: rated voltage $V_n = 220$ V, rated armature-current $I_{a,n} = 100$ A, rated armature resistance $r_a = 0.05$ Ω, and rated shaft-speed of 970 rpm.

   When the motor rotates at $n_1 = 1000$ rpm, a plugging braking mode is applied to bring the drive to a standstill (figure P4.2).

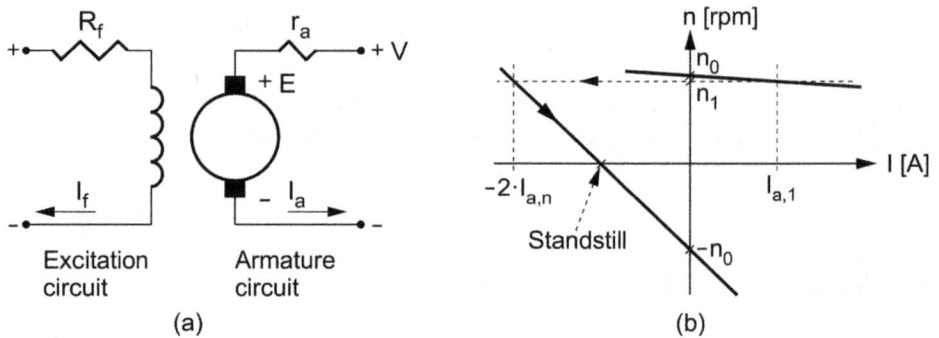

Figure P4.2. Plugging braking mode of a DC motor. (a) Circuit diagram, and (b) speed versus current curves.

   Calculate:
   (a) The external resistance to be connected in the armature to limit the braking current to twice the rated armature-current.
   (b) The value of the braking torque at $n_1 = 1000$ rpm.
   (c) The value of the EM torque at standstill.

3. A shunt-connected DC motor has the following rated parameters: $P_n = 20$ kW, $V_n = 220$ V, $n_n = 1200$ rpm, $\eta_n = 0.88$, and an armature resistance $r_a = 0.1$ Ω. It operates at rated torque T_n and rated speed ω_n. To stop the drive, the armature was reverse-switched abruptly; that is, the voltage across the armature was reversed, but not across its excitation winding (figure 4.8). An external resistor R_E was inserted in the armature circuit to limit the maximum braking torque to $T_b = -2T_{em,\,n}$. The speed–torque characteristics of the event is presented (figure P4.3)

   Assume that the resistive torque referred to the motor-shaft is constant at rated torque value, and neglect the excitation current. Address the following:
   (a) Calculate the value of the external resistor in the armature circuit.
   (b) Show (write) the equation of the speed versus torque during the plugging mode.

(c) Calculate the stoppage time of the drive, from rated speed to standstill, knowing that the equivalent moment of inertia referred to the motor shaft is $J = 1.0$ Nm s^2 (see section 3.1).

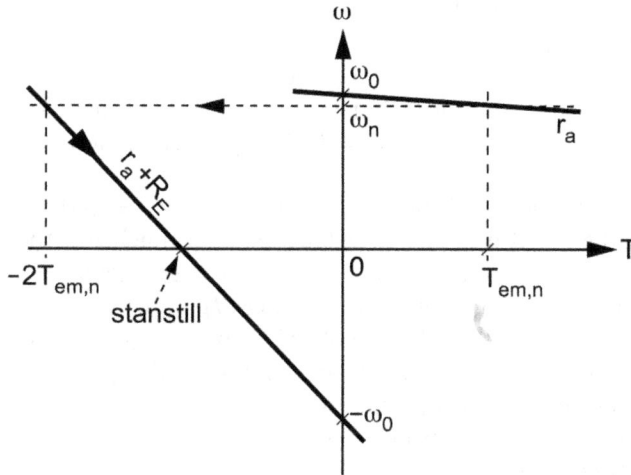

Figure P4.3. Plugging braking mode to reach a complete stop.

4. The motor in question #3 operates at rated torque T_n and rated speed ω_n. To stop the drive, a dynamic braking mode (figure 4.7) is applied. An external resistor R_E was inserted in the armature circuit to limit the maximum braking torque to $T_b = -2T_{em,\, n}$. The speed–torque characteristics of the event is presented (figure P4.4).

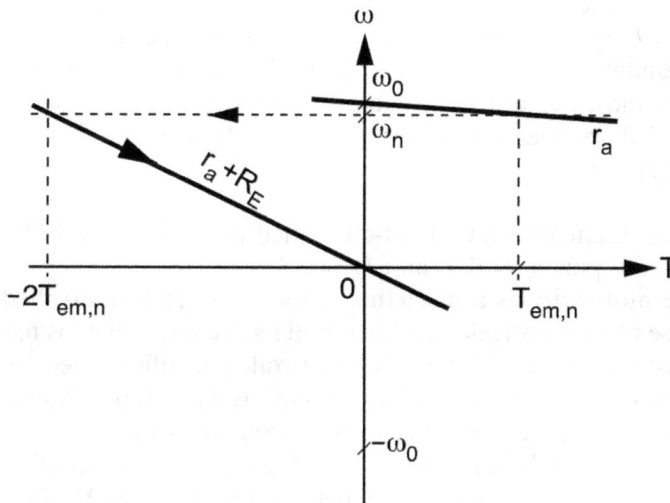

Figure P4.4. Dynamic braking mode to reach a complete stop.

Assume that the resistive torque referred to the motor-shaft is constant at rated torque value, and neglect the excitation current. Address the following:
   (a) Calculate the value of the external resistor in the armature circuit.
   (b) Show (write) the equation of the speed versus torque during the plugging mode.
   (c) Calculate the stoppage time of the drive, from rated speed to standstill, knowing that the equivalent moment of inertia referred to the motor shaft is $J = 1.0$ Nm s^2 (see section 3.1).

5. A shunt-connected DC motor: $P_n = 1.7$ kW, $V_n = 200$ V, $n_n = 1620$ rpm, and $r_a = 0.62$ Ω, drives a fan. The fan is directly attached to the motor shaft (w/o a gearbox), and the total moment of inertia at the motor shaft is $J = 0.2$ Nm s^2. The torque-speed characteristic of the fan is: $T_R = 0.5 + 3.3 \cdot 10^{-4} \cdot \omega^2$ [Nm], The drive operates at rated speed of 1620 rpm (first quadrant operation).
   At $t = 0$, the voltage across the armature is abruptly reversed (a plugging braking mode) to bring the fan to a complete stop. To limit the current during the braking mode, a resistor $R_E = 44.55$ Ω was inserted in the armature circuit.
   Neglect all losses, and address the following:
   (a) Draw a qualitative graph $\omega = f(T)$ of four curves: the motor natural curve, the load (fan) curve, the motor plugging mode curve, and the dynamic torque curve during the brake.
   (b) Calculate the deceleration time from full-speed to standstill (see section 3.1).

6. The speed of a DC shunt-connected motor is regulated by a 6-pulse phase-controlled rectifier. The motor nameplate states: 3.24 kW, 120 V, 30 A, and 970 rpm. The rectifier itself is fed from a 3-ϕ symmetrical system at 120 V phase-voltage via a power transformer whose leakage reactance referred to its secondary winding is 0.2 Ω/ph. Neglect the excitation current of the DC motor, and calculate the value of the delayed firing angle α of the phase-controlled rectifier to achieve a motor shaft-speed of 500 rpm at rated motor current.

7. The nameplate of a DC shunt-connected motor is: $P_n = 5$ kW, $V_n = 200$ V, $n_n = 1200$ rpm, $\eta_n = 0.9$, and $I_f = 2$ A.
   The motor drives a mechanism. Its armature is controlled by a 3-ϕ, 6-pulse phase-controlled rectifier, and its excitation field is fed separately at constant current of 2 A. The controlled rectifier is fed by a 3-ϕ transformer, where its secondary voltage is $V_2 = 160$ V line-to-line and its leakage reactance referred to the secondary winding is: $x_c = 0.2$ Ω/ph. At a phase-delayed firing angle of the controlled rectifier $\alpha = 30°$, the motor develops an electromagnetic torque of $T_{em} = 25$ Nm. Calculate the motor shaft-speed.

8. The nameplate of a DC shunt-connected motor is: 5.1 kW, 200 V, 30 A, and 890 rpm. Its shaft speed is regulated by a 3-ϕ, 6-pulse phase-controlled rectifier. The rectifier itself is fed from a 3-ϕ symmetrical system at 460 V line-to-line. The commutating reactance per-phase of the transformer seen at its secondary winding is $X_c = 0.3\ \Omega/\text{ph}$.

   Neglect the excitation current of the DC motor, and calculate the value of the delayed firing angle α of the phase-controlled rectifier that is required to achieve motor shaft-speed of 350 rpm at rated motor current.

9. A 3-ϕ, 6-pulse, AC/DC phase-controlled rectifier supplies variable DC voltage to the armature of a separately excited DC motor. The excitation field is fed separately at constant current.

   A 3-ϕ transformer feeds the controlled rectifier (figure P4.9).

   The transformer primary voltage (the utility voltage) is: $V_1 = 480$ V line-to-line, and its leakage reactance referred to the secondary winding is: $x_c = 0.3\ \Omega/\text{ph}$.

   The motor: nameplate states: 5 kW, 216 V, 30 A, and 1000 rpm.

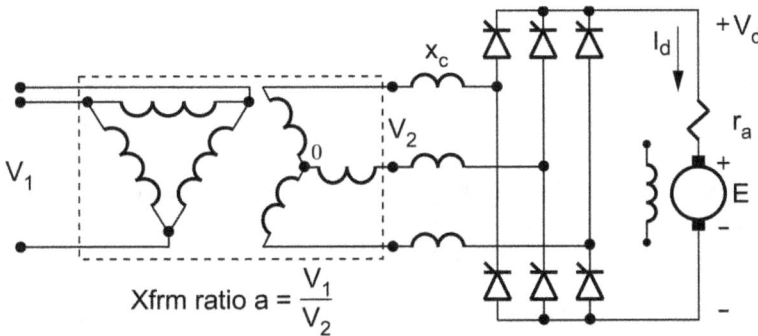

Figure P4.9. A 3-ϕ, 6-pulse, phase-controlled rectifier controls the speed of a DC motor.

   Neglect the excitation current and address the following questions:
   (a) What is the transformer voltage ratio.
   (b) What is the range of the phase-delayed firing angle, $\alpha_{min} \leqslant \alpha < \alpha_{mx}$, to allow a speed control range of $900 \geqslant n > 0$ rpm at rated current.
   (c) What is the range of armature voltage to accommodate the speed range in (b)?

10. A separately excited DC motor has the following nameplate data: 5 kW, 200 V, 27.8 A, and 900 rpm. A 3-ϕ, 6-pulse, phase-controlled rectifier regulates the voltage across the motor armature (figure P4.10).

    The 3-ϕ transformer that feeds the rectifier has a rated power of 5 kVA, a voltage ratio $V_1/V_2 = \frac{416}{208\ \text{V}}$, and a per-unit impedance of 0.04 p. u.

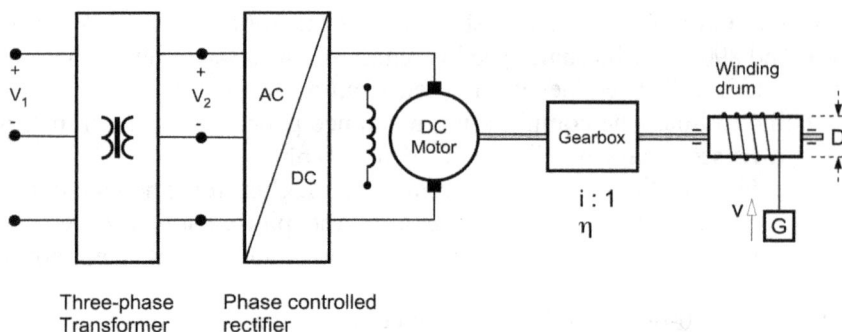

Figure P4.10. Voltage regulated DC motor drives a winding drum.

The motor drives a winding drum through a gearbox having a gear ratio $i = 6$, and an efficiency 90%. The drum diameter is $D = 0.6$ m, and it lifts a load $G = 90$ kg$_f$ at a constant velocity $v = 3$ m s^{-1}.

Neglect the excitation current of the motor, and assume that the EM torque equals the shaft torque. Calculate the delayed firing angle α of the phase-controlled rectifier.

References

[1] Fitzgerald A E, Kingsley C and Umans S 2002 *Electric Machinery* 6th edn (New York: McGraw-Hill)

[2] Kostenko M and Piotrovsky L 1974 *Electrical Machines* **2 volumes** 3rd edn (Moscow: MIR Publishers)

[3] Wildi T 2006 *Electrical Machines, Drive and Power Systems* 6th edn (Englewood Cliffs, NJ: Prentice-Hall)

[4] Mohan N 2001 *Electric Drives, An Integrative Approach* (MNPERE Publisher)

[5] Dubey G K 2001 *Fundamentals of Electrical Drives* (Alpha Science International Ltd)

[6] Nasar S A and Unnewehr L E 1983 *Electromechanical and Electric Machines* (New York: Wiley)

[7] Meyers R A (ed) 2002 *Encyclopedia of Physical Science and Technology* **vol 5** 3rd edn (New York: Academic)

[8] Chilikin M 1976 *Electric Drive* (Moscow: MIR Publishers)

[9] Erickson R W and Maksimović D 2020 *Fundamentals of Power Electronics* 3rd edn (Berlin: Springer)

[10] Schaefer J 1965 *Rectifier Circuits: Theory and Design* (New York: Wiley)

[11] Rashid M H 2004 *Power Electronics: Circuits, Devices, and Applications* 3rd edn (Englewood Cliffs, NJ: Pearson Prentice-Hall)

[12] Csaki F, Ganszky K, Ipsits I and Marti S 1975 *Power Electronics* (Gudapest: Akademiai Kiado)

[13] Zabar Z 2022 *Fundamentals of Distributed Generation Systems* (New York: AIP Publishing)

IOP Publishing

Fundamentals of Electromechanical Drives

Zivan Zabar

Chapter 5

The induction (asynchronous) motor

The induction motor is the most common type of industrial electric motors. It is a reliable, efficient, relatively light weight at low cost, and requires minimal maintenance. It finds applications in elevators, air conditioners, cranes, hoists, refrigeration systems, pumps and more. This chapter addresses briefly its principle of operation, its power and torque relationships, its braking modes of operation, and its speed control techniques.

5.1 Principle of operation

The operating principle of an induction motor is illustrated below (figure 5.1). Two magnetic poles, North and South, generate the main magnetic flux (green arrow). Assuming, for the sake of argument, that those two magnetic poles rotate concurrently in the counter-clockwise direction at an angular speed ω [s^{-1}]. This motion causes the main magnetic field in the airgap, between the poles and the

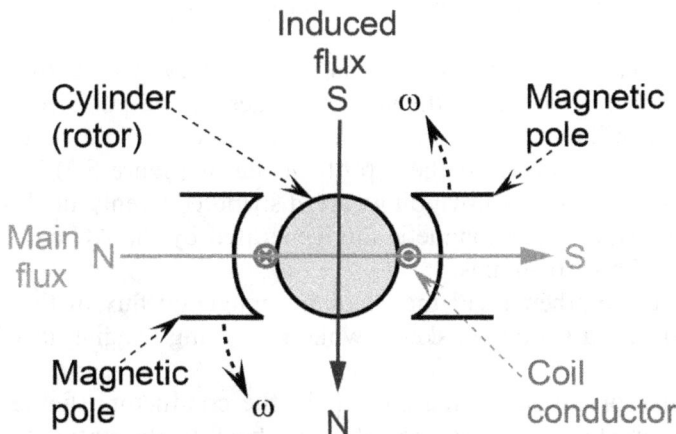

Figure 5.1. Principle-operation of an induction motor.

cylinder, to rotate at the same speed. The cylinder itself is made of laminated ferromagnetic materials to minimize iron losses due to eddy currents.

A conducting coil is wound securely along the length of that cylinder (red circles in figure 5.1) and moves with it. According to Faraday's law (equation (4.4)), current is induced (induction motor) in the two coil conductors and flows in the direction shown in the figure. That induced current generates its own magnetic field (blue arrow) perpendicular to the main flux.

Because same poles repel and opposite poles attract, the cylinder will rotate in the counter-clockwise direction too. In short, the rotating magnetic field drags the cylinder (the rotor) in its direction of motion.

In reality, the poles are stationary. Nevertheless, they can generate a rotating magnetic field in the airgap of the motor at the frequency of the input current. The brilliant idea of generating a rotating magnetic field from a set of stationary poles, was developed by Nicola Tesla at the end of the 19th century. The following subsections present a mathematical model for the rotating magnetic field [1–3].

5.1.1 Mathematical model of a stationary magnetic wave

In essence, a 1-ϕ, single-coil induction motor is shown below (figure 5.2).

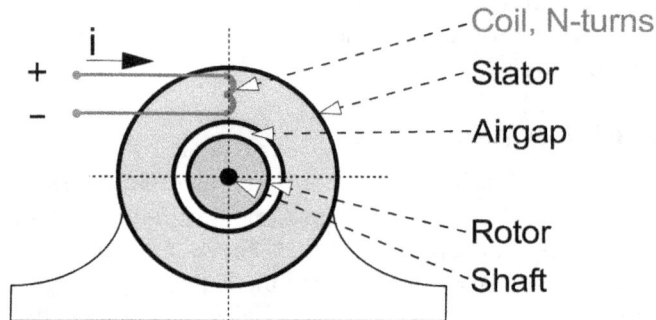

Figure 5.2. An elementary single-phase induction motor.

The motor consists of two main parts, a stationary part called stator, and a rotating part called rotor where the shaft provides the output torque. The airgap, usually of a few millimeters, allows the rotor to rotate without rubbing the stator. The coil with N-turns shown at the top of the stator (figure 5.2) is fed by an AC current i. In practice, the coil-windings are distributed evenly in slots around the periphery of the stator. The magnetic flux generated by the AC current in the coil oscillates along the vertical axis.

To construct a mathematical model of the magnetic flux in the airgap of the motor, one can use a simplified sketch while neglecting fringing flux in the airgap (figure 5.3).

The flux lines produced by the current in the conductors (figure 5.3), present closed loops from stator to airgap to rotor and back to the stator. For clarity, one

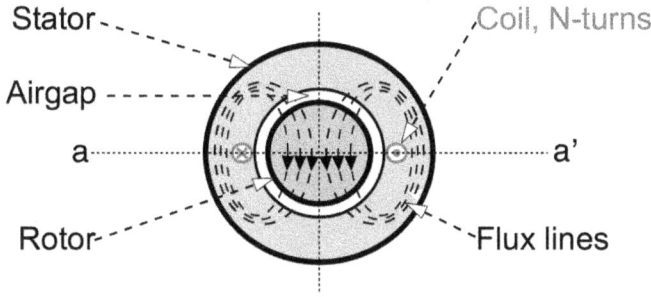

Figure 5.3. Flux lines produced by the N-turn coil for a given current.

can dissect the motor along the line a–a' and flat open the two parts. The linear presentation of those magnetic flux lines would be (figure 5.4).

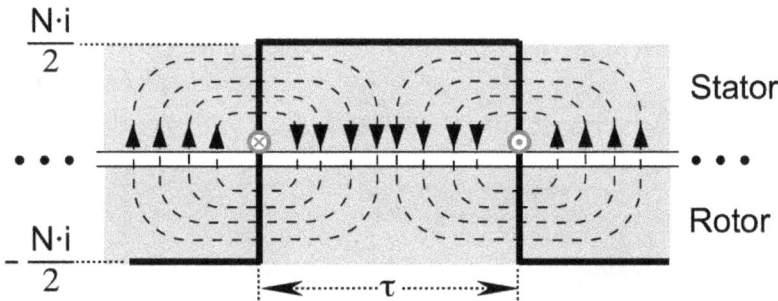

Figure 5.4. Linear presentation of the flux distribution along the airgap for a given current.

The pole pitch τ is the peripheral distance between two adjacent pole centers (figure 5.4). Clearly, in a two-pole machine (figures 5.2–5.4), there are two pole-pitches. The product $N \cdot i$ presents the ampere-turns per-pole.

Assuming constant ampere-turns under each pole pitch, the distribution of the flux lines along the perimeter of the airgap resembles a cyclic square-wave function where the period is $2 \cdot \tau$ (figure 5.4). The Fourier series of that wave is:

$$F(x, t) = \frac{4}{\pi} \cdot \frac{Ni}{2} \sum_n \cos n\omega_x \cdot x = \frac{4}{\pi} \cdot \frac{Ni}{2} \sum_n \cos n\frac{\pi x}{\tau} \tag{5.1}$$

where

$i = I_{mx} \sin \omega t$ [A] is the current in the coil,
N is the number of turns in the coil,
n is the harmonic number,
ω_x The radial or geometric (space) frequency,
x [m] is the distance along the periphery of the airgap, and
τ [m] is the pole pitch.

Consider $p \geqslant 2$, one complete revolution is:

$$\pi \cdot D = p \cdot \tau \implies \tau = \frac{\pi \cdot D}{p} \ [\text{m}] \tag{5.2}$$

where

D [m] is the diameter of the rotor and
p is the number of poles

In practice, the coil winding is distributed evenly around the periphery of the stator, such that a sinewave distribution of magnetic flux in the airgap can actually be considered (not a square-wave distribution). For that reason, only the fundamental component of $F(x, t)$ function (equation (5.1)) can be used:

$$F_1(x, t) = \frac{4}{\pi} \frac{Ni}{2} \cdot \cos \frac{\pi}{\tau} x \tag{5.3}$$

Substituting the current $i = I_{mx} \sin \omega t$ in equation (5.3) gives:

$$\left. \begin{array}{c} F_1(x, t) = \dfrac{4}{\pi} \dfrac{NI_{mx}}{2} \cdot \cos \dfrac{\pi}{\tau} x \cdot \sin \omega t = F_{mx} \cdot \cos \dfrac{\pi}{\tau} x \cdot \sin \omega t \\[2mm] \text{where } F_{mx} = \dfrac{4}{\pi} \dfrac{NI_{mx}}{2} \end{array} \right\} \tag{5.4}$$

For instance, the ampere-turns at two time instances, $\omega t_1 = \pi/2$ and $\omega t_6 = -\pi/2$, results in (table 5.1).

Table 5.1. Ampere-turns at two different time instances, ωt_1 and $\omega \, t_6$ (equation (5.4)).

ωt_1	$\pi/2$	$\pi/2$	$\pi/2$	$\pi/2$	$\pi/2$
Position x	$-\tau$	$-\tau/2$	0	$\tau/2$	τ
Ampere-turns	$-F_{mx}$	0	F_{mx}	0	$-F_{mx}$
ωt_6	$-\pi/2$	$-\pi/2$	$-\pi/2$	$-\pi/2$	$-\pi/2$
Position x	$-\tau$	$-\tau/2$	0	$\tau/2$	τ
Ampere-turns	F_{mx}	0	$-F_{mx}$	0	F_{mx}

The behavior of the ampere-turns along the periphery of the airgap, $F_1(x, t)$, at six time instances is shown in figure 5.5.

As shown (figure 5.5), the magnetic wave does oscillate in time but its amplitude profiles do not move in space (position). That behavior is termed 'a stationary wave' also called 'a standing wave.'

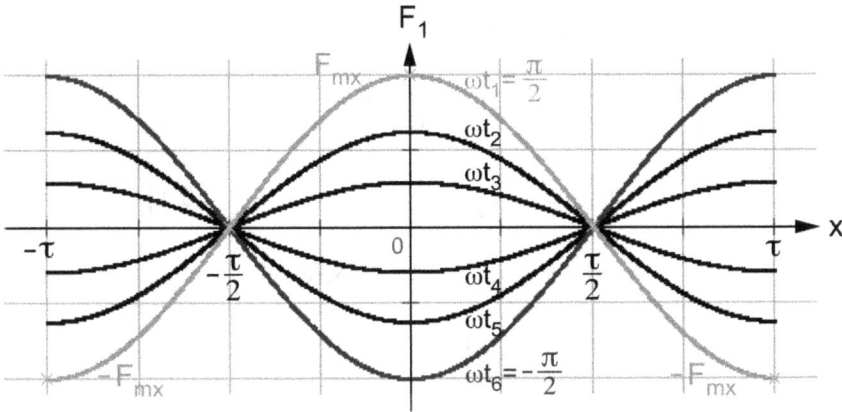

Figure 5.5. 1-ϕ flux distribution along the periphery of the airgap at different time instances.

5.1.2 Mathematical model of a travelling magnetic wave

In a 3-ϕ AC motor, three independent coils are positioned around the stator. Those coils are spaced evenly at 120° apart (figure 5.6).

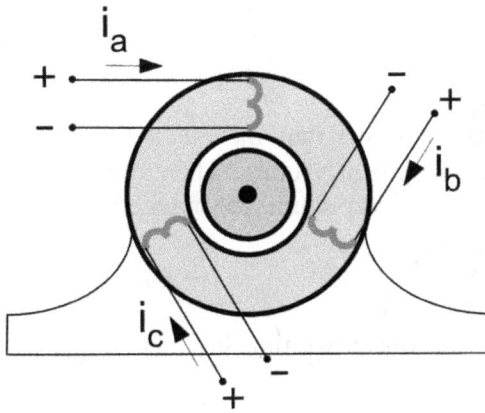

Figure 5.6. An elementary 3-ϕ induction motor.

The three coils are fed by a symmetrical set of 3-ϕ voltages that, in practice, generate a set of 3-ϕ currents:

$$\left. \begin{array}{l} i_a = I_{mx} \sin(\omega t) \\ i_b = I_{mx} \sin(\omega t - 120°) \\ i_b = I_{mx} \sin(\omega t + 120°) \end{array} \right\} \tag{5.5}$$

In practice, the windings of the three coils are distributed evenly in slots around the periphery of the stator. Each coil generates its own standing magnetic wave in the airgap of the motor (see equation (5.4)):

$$F_a(x, t) = F_{mx} \cos\left(\frac{\pi}{\tau}x\right) \cdot \sin(\omega t)$$

$$F_b(x, t) = F_{mx} \cos\left(\frac{\pi}{\tau}x - 120°\right) \cdot \sin(\omega t - 120°) \right\} \quad (5.6)$$

$$F_c(x, t) = F_{mx} \cos\left(\frac{\pi}{\tau}x + 120°\right) \cdot \sin(\omega t + 120°)$$

where $F_a(x, t)$, $F_b(x, t)$, and $F_c(x, t)$ are the ampere-turns generated by each of the three stator-coils. The combined magnetic flux in the airgap would be the contribution (algebraic summation) of those three coils:

$$F_T(x, t) = F_a(x, t) + F_b(x, t) + F_c(x, t) \quad (5.7)$$

Using the trigonometric identity [5]: $\sin\alpha \cdot \cos\beta = \frac{1}{2}[\sin(\alpha - \beta) + \sin(\alpha + \beta)]$, one gets:

$$F_T(x, t) =$$

$$= \frac{F_{mx}}{2}\left[\begin{array}{l} \sin\left(\omega t - \frac{\pi x}{\tau}\right) + \sin\left(\omega t + \frac{\pi x}{\tau}\right) \\ +\sin\left(\omega t - 120° - \frac{\pi x}{\tau} + 120°\right) + \sin\left(\omega t - 120° + \frac{\pi x}{\tau} - 120°\right) \\ +\sin\left(\omega t + 120° - \frac{\pi x}{\tau} - 120°\right) + \sin\left(\omega t + 120° + \frac{\pi x}{\tau} + 120°\right) \end{array}\right] \right\} \quad (5.8)$$

The three terms (equation (5.8)) that include the parameter $(\omega t + \frac{\pi x}{\tau})$ create a balanced (symmetrical) system of vectors. Therefore, their summation is nullified (zero), and the final formula that presents the magnetic flux behavior in the airgap is:

$$F_T(x, t) = \frac{3F_{mx}}{2}\sin\left(\omega t - \frac{\pi x}{\tau}\right) = F'_{mx}\sin\left(\omega t - \frac{\pi x}{\tau}\right) \quad (5.9)$$

For instance, the ampere-turns at two time instances, $\omega t_1 = 0$ and $\omega t_4 = \pi$, results in table 5.2.

Table 5.2. Ampere-turns at two different time instances, ωt_1 and ωt_4 (equation (5.9)).

ωt_1	0	0	0	0	0
Position x	$-\tau$	$-\tau/2$	0	$\tau/2$	τ
Ampere-turns	0	F_{mx}	0	$-F_{mx}$	0

ωt_4	π	π	π	π	π
Position x	$-\tau$	$-\tau/2$	0	$\tau/2$	τ
Ampere-turns	0	$-F_{mx}$	0	F_{mx}	0

The behavior of the ampere-turns along the periphery of the airgap, $F_T(x, t)$, at four time instances is shown in figure 5.7.

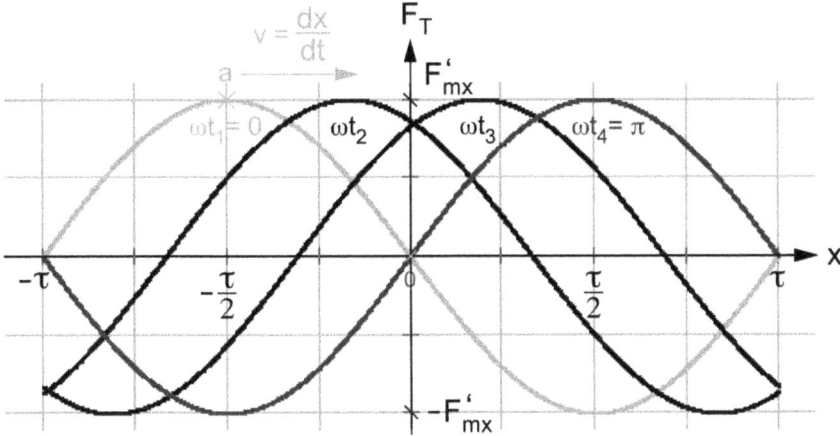

Figure 5.7. 3-ϕ flux distribution along the periphery of the airgap at different time instances.

- Consider the peak amplitude F'_{mx} at point a (figure 5.7). As time progresses, that peak moves (travels) along the periphery of the airgap at a linear speed of v [m s^{-1}]. In essence, the entire wave travels around the machine airgap at the linear speed v, which suggests the term 'travelling wave'. In short, the combined magnetic flux (equation (5.7)), which is the main magnetic flux, rotates around the periphery of the airgap.
- Consider again the peak amplitude F'_{mx} at point a (figure 5.7). As the wave travels at a linear speed v, that peak amplitude remains constant, $F_T(x, t) = $ const. (equation (5.9)), that is:

$$F'_{mx} \sin\left(\omega t - \frac{\pi x}{\tau}\right) = \text{const.} \implies \left(\omega t - \frac{\pi x}{\tau}\right) = \text{const.} \tag{5.10}$$

The derivation of equation (5.10) with respect to time suggests:

$$\left.\begin{array}{c} \omega \cdot dt - \dfrac{\pi}{\tau} \cdot dx = 0 \\ \text{or} \\ \dfrac{dx}{dt} = \dfrac{\omega \cdot \tau}{\pi} \end{array}\right\} \tag{5.11}$$

And the *linear speed* v of the magnetic wave is:

$$v = \frac{dx}{dt} = \frac{2\pi \cdot f \cdot \tau}{\pi} = 2\tau \cdot f \quad [\text{m s}^{-1}] \tag{5.12}$$

where f [s^{-1}] is the frequency of the input current.

- Consider once again the peak amplitude F'_{mx} at point a (figure 5.7). In one minute, that peak travels along the circumference of the machine:

$$v \cdot 60 = p \cdot \tau \cdot n \text{ [m]} \tag{5.13}$$

where

v [m s^{-1}] is the linear speed (equation (5.12)) of the peak amplitude F'_{mx},
p is the number of magnetic poles,
τ [m] is the pole pitch, and
n [rpm] is the rotational speed in revolutions per minute.

And the rotational speed of the magnetic wave is:

$$n = \frac{v \cdot 60}{p \cdot \tau} = \frac{2\tau \cdot f \cdot 60}{p \cdot \tau} = \frac{120 \cdot f}{p} \text{ [rpm]} \tag{5.14}$$

where

f [s^{-1}] is the frequency of the input current, and
p is the number of magnetic poles.

5.1.3 Action of the rotor

The rotor of the induction motor is a cylinder made of laminated sheets of ferromagnetic material having slots for the copper or aluminum conductors (figure 5.8).

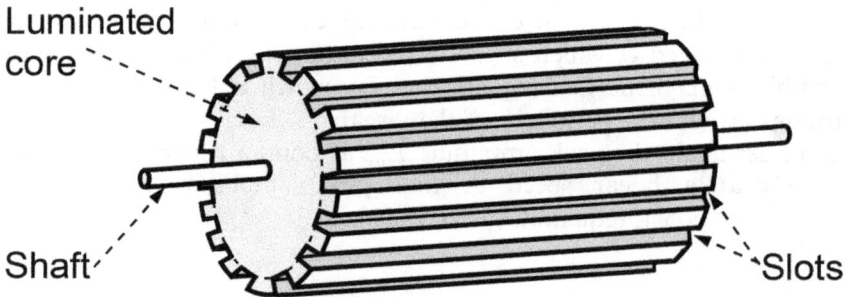

Figure 5.8. 3D view of the rotor of an induction motor, without conductors.

Two types of coil winding are available in the rotor:

- *Squirrel-cage rotor*—the conductors (usually conducting bars) are laid out in the slots and shorted together at both ends of the rotor by conducting end-rings. The advantages are low maintenance requirements, high reliability, and relatively low cost.
- *Wound rotor*—the conductors are laid out in the slots in a 3-ϕ formation and connected to a set of three slip rings. The advantage is the ability to providing high level of control in speed and in torque.

At start (rotor at standstill), the rotating magnetic field in the airgap induces voltages in the rotor winding at input frequency. If the winding conductors are short-circuited, the electric currents would induce magnetic flux in the rotor. The interaction between the main flux and the induced flux generates a starting torque (same as in figure 5.1). As the speed increases, the induced currents and the generated torque will also vary (see section 5.3). If the rotor rotates at the speed of the rotating magnetic field (synchronized with it), there would not be any magnetic flux change and eventually zero induced currents in the rotor. In short, there must be a difference (slip) in speed between the rotating magnetic field in the airgap and the moving rotor for the induction motor to develop a torque. That speed difference suggests another name for the motor, which is: 'asynchronous motor.'

5.1.4 Definition of slip

As stated above, the rotor speed lags behind the speed of the rotating magnetic wave. Both speeds are portrayed by arrows in figure 5.9.

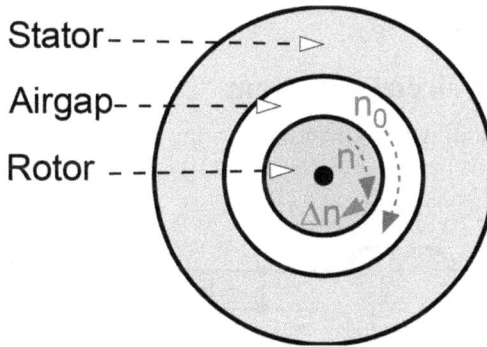

Figure 5.9. Speed relations.

By definition, the difference Δn between the two speeds (figure 5.9) is:

$$\Delta n = n_0 - n \quad [\text{rpm}] \tag{5.15}$$

Slip is defined as the normalized value of the speed difference:

$$S = \frac{\Delta n}{n_0} = \frac{n_0 - n}{n_0} = \frac{\omega_0 - \omega}{\omega_0} \tag{5.16}$$

where
   S is the slip,
   n_0 [rpm] is termed the synchronous speed of the motor, which is the speed of the rotating magnetic field (equation (5.14)), and
   n [rpm] is the actual rotational speed of the rotor, and $\omega = 2\pi \cdot n/60 \ [\text{s}^{-1}]$ is the radial speed.

Multiplying both side of equation (5.15) by the ratio $(p/120)$,

$$\frac{p}{120}\Delta n = \frac{p}{120}n_0 - \frac{p}{120}n \implies f_2 = f_1 - f_r \ [\text{s}^{-1}] \tag{5.17}$$

suggests that the frequency f_2 of the induced currents in the rotor is the difference between the stator frequency f_1 and the mechanical (rotational) frequency f_r of the rotor. Equation (5.17) implies that although the mechanical speed of the rotor is different than the synchronous speed, the resultant rotating magnetic flux generated by the rotor itself is actually synchronized with the rotating magnetic flux generated by the stator.

Using the slip definition (equation (5.16)), the rotor frequency f_2 can also be expressed in terms of the stator frequency f_1:

$$f_2 = \frac{p \cdot \Delta n}{120} = \frac{p \cdot S n_0}{120} = S \cdot f_1 \ [\text{s}^{-1}] \tag{5.18}$$

The equation above (equation (5.18)) indicates that the frequency f_2 of the currents induced in the rotor depend on the rotor speed. For instance, at standstill, $n = 0$ or $S = 1$, the frequency of the currents in the rotor equals the frequency of the currents in the stator. At full speed, the motor slip is only a few percentage points, which suggests that the frequency of the currents in the rotor is relatively very small.

5.2 The equivalent electric diagram

The equivalent electric diagram of induction motor is [1–3]:

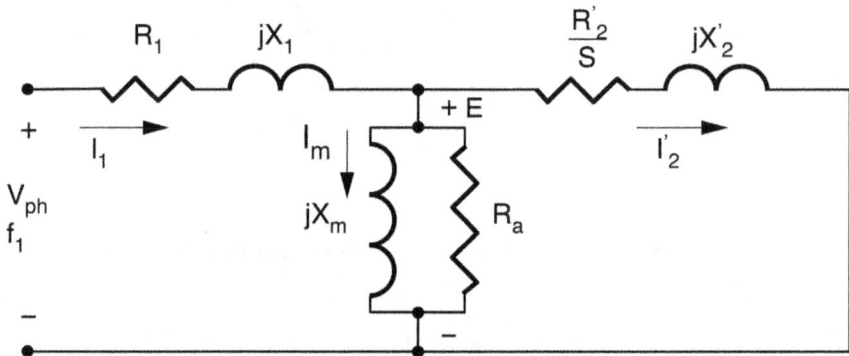

Figure 5.10. The equivalent diagram per-phase of an induction motor.

In figure 5.10:
R_1 [Ω/ph] is the stator resistance per phase,
X_1 [Ω/ph] is the leakage reactance per phase in the stator,
E [V/ph] is the electromotive force (EMF) per phase in the airgap of the motor,
X_m [Ω/ph] is the magnetizing reactance per phase,
I_m [A] is the magnetizing current per phase,
R_a [Ω/ph] is an equivalent resistance related to the core losses,
R_2' [Ω/ph] is the rotor resistance per phase referred to the stator,
X_2' [Ω/ph] is the leakage reactance per phase in the rotor referred to the stator,
S is the slip (equation (5.16)),
I_1 [A] is the phase current in the stator,
I_2' [A] is the phase current in the rotor referred to the stator,
V_{ph} [V/ph] is the input voltage per phase, and
f_1 [s^{-1}] is the frequency of the input voltage.

Note that $R_2' = \alpha^2 R_2$, $X_2' = \alpha^2 X_2$, and $I_2' = \frac{I_2}{\alpha}$;

where $\alpha \cong \frac{N_1}{N_2} = \frac{\text{Number of turns per phase in stator}}{\text{Number of turns per phase in rotor}}$.

In practice, the core losses are minimal and the resistance R_a can be neglected. Further simplification allows for the magnetizing reactance to be inserted at the input terminals of the motor [1–3].

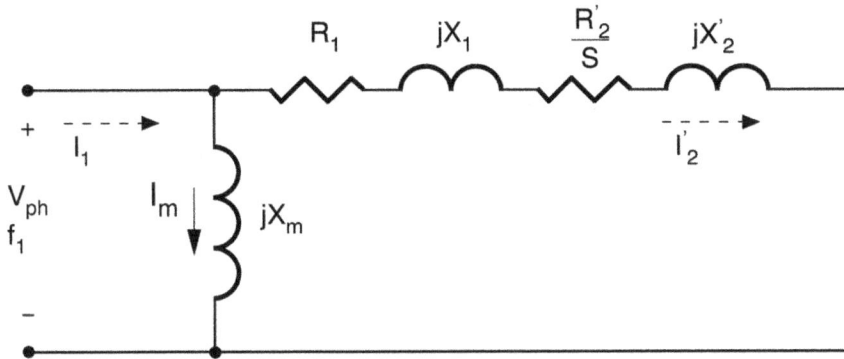

Figure 5.11. The simplified equivalent diagram per phase of an induction motor.

The simplified diagram (figure 5.11) has two important objectives: first, it facilitates the calculation of the rotor current I_2', which generates the shaft torque (section 5.3). Second, the assumption that the voltage across the magnetizing reactance equals the input voltage, $E = V_{ph}$, emphasizes the fact that the required magnetizing current I_m is supplied by the utility line.

5.3 Power distribution and torque development

The distribution of active power in an induction motor is shown in figure 5.12.

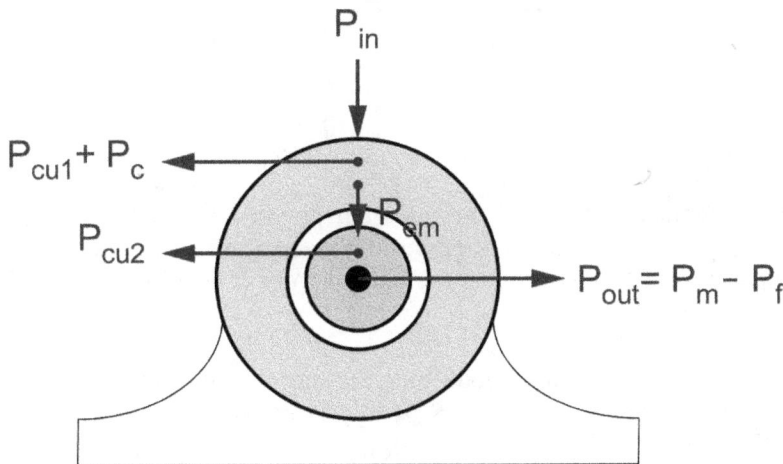

Figure 5.12. Power distribution in an induction motor.

In figure 5.12:

P_{in} [W] is the input (utility) power to the motor,

P_{cu1} [W] is the copper losses in the stator,

P_c [W] is the core losses (hysteresis and eddy current losses),

P_{em} [W] is the electromagnetic (EM) power in the airgap,

P_{cu2} [W] is the copper losses in the rotor,

P_{out} [W] is the shaft power,

P_m [W] is the mechanical power developed by the motor, and

P_f [W] is the friction losses (bearing and viscous friction).

In practice, the friction losses are minimal and can be neglected. That is:

$$P_{out} \cong P_m \ [W] \tag{5.19}$$

The mechanical power is:

$$P_m = P_{em} - P_{cu2} \ [W] \tag{5.20}$$

The copper losses in the rotor are (see figure 5.10):

$$P_{cu2} = 3 \cdot \left(I_2'\right)^2 \cdot R_2' \ [W] \tag{5.21}$$

The EM power in the airgap is (see figure 5.10):

$$P_{em} = 3 \cdot \left(I_2'\right)^2 \frac{R_2'}{S} = \frac{P_{cu2}}{S} \ [W] \tag{5.22}$$

Substitute P_{cu2} (equation (5.22)) in equation (5.20) suggests:

$$P_m = P_{em} - S \cdot P_{em} = P_{em}(1 - S) \ [W] \tag{5.23}$$

Note: The resistance R_2 of the rotor is usually not available in the data sheet of the motor. For a wound rotor, it can be estimated straight from the motor nameplate. Using equation (5.22):

$$\left. \begin{aligned} P_{cu2,\ n} &= S_n \cdot P_{em,\ n} \\ \text{or: } 3 \cdot (I_{2n})^2 \cdot R_2 &\cong S_n \cdot \sqrt{3} \ E_{2n} I_{2n} \\ \text{Finally: } R_2 &\cong \frac{S_n \cdot E_{2n}}{\sqrt{3} \ I_{2n}} \end{aligned} \right\} \tag{5.24}$$

where

R_2 [Ω/ph] is the resistance of the rotor per phase,

E_{2n} [V] is the open-circuit voltage of the wound rotor (given in the nameplate),

I_{2n} [A] is the nominal current in the wound rotor (given in the nameplate), and

n denotes the rated value.

Note: Under normal operating conditions, the motor slip is only a few percentage points, which suggests that $R_2'/S \gg X_2'$ (see figure 5.10). As a result, $P_{em,\ n} \cong \sqrt{3} \ E_{2n} I_{2n}$ is a practical assumption.

The EM torque T developed in the airgap can be determined from:

$$T = \frac{P_{em}}{\omega_0} = \frac{3}{\omega_0}(I_2')^2 \frac{R_2'}{S} \ \text{[Nm]} \tag{5.25}$$

where the synchronous angular speed ω_0 is:

$$\omega_0 = \frac{2\pi \cdot n_0}{60} \ \text{[s}^{-1}\text{]} \tag{5.26}$$

And the current I_2' can be calculated from the simplified equivalent diagram (figure 5.11):

$$I_2' = \frac{V_{ph}}{\sqrt{\left(R_1 + \dfrac{R_2'}{S}\right)^2 + \left(X_1 + X_2'\right)^2}} \ \text{[A]} \tag{5.27}$$

By substituting equation (5.27) into equation (5.25), the EM torque equation becomes:

$$T = \frac{3}{\omega_0} \frac{V_{ph}^2}{\left[\left(R_1 + \dfrac{R_2'}{S}\right)^2 + \left(X_1 + X_2'\right)^2\right]} \frac{R_2'}{S} \ \text{[Nm]} \tag{5.28}$$

The EM torque versus slip curve, $T = f(S)$, of the machine is presented in figure 5.13.

The shaft speed versus the EM torque curve, $\omega = f(T)$, of the machine would be as in figure 5.14.

The curves (figures 5.13 and 5.14) present three modes of operation:

- *Motoring* mode, where $1 \geqslant S > 0$, or $0 \leqslant \omega < \omega_0$.
- *Braking* mode, where $S > 1$, or $\omega < 0$.
- *Generating* mode, where $S < 0$, or $\omega > \omega_0$.

The value of maximum torque T_{mx} at the critical slip and the minimum torque T_{min} (figures 5.13 and 5.14) can be obtained from equation (5.28). The result of the derivation is [1–3]:

- *The critical slip S_C for those peak values is:*

$$S_C = \frac{R_2'}{\pm\sqrt{R_1^2 + \left(X_1 + X_2'\right)^2}} \cong \pm\frac{R_2'}{X_1 + X_2'} \tag{5.29}$$

The approximation above (equation (5.29)) relates to the fact that in practice: $R_1 \ll X_1 + X_2'$.
- *The magnitude of the peak EM torque* can be obtained by substituting S_C (equation (5.29)) into equation (5.28):

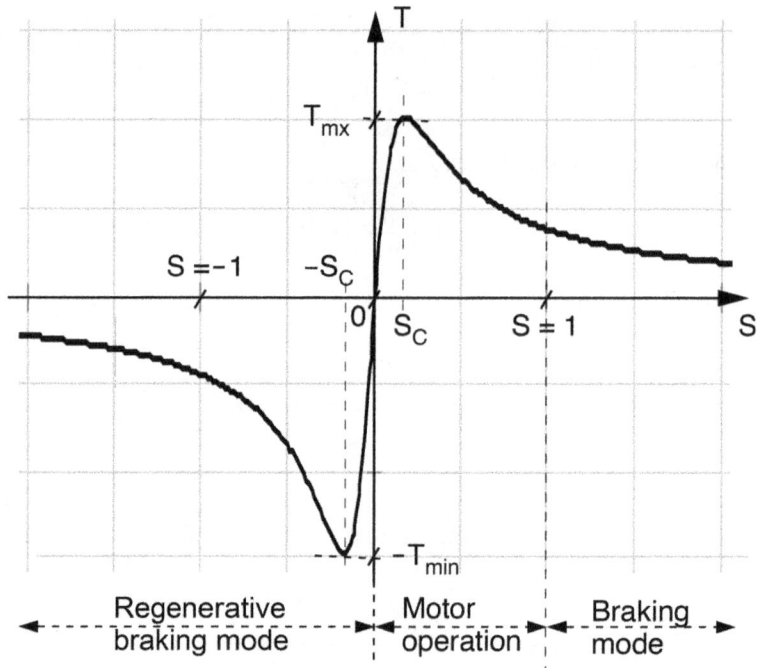

Figure 5.13. EM torque versus slip characteristic, $T = f(S)$, of an induction machine.

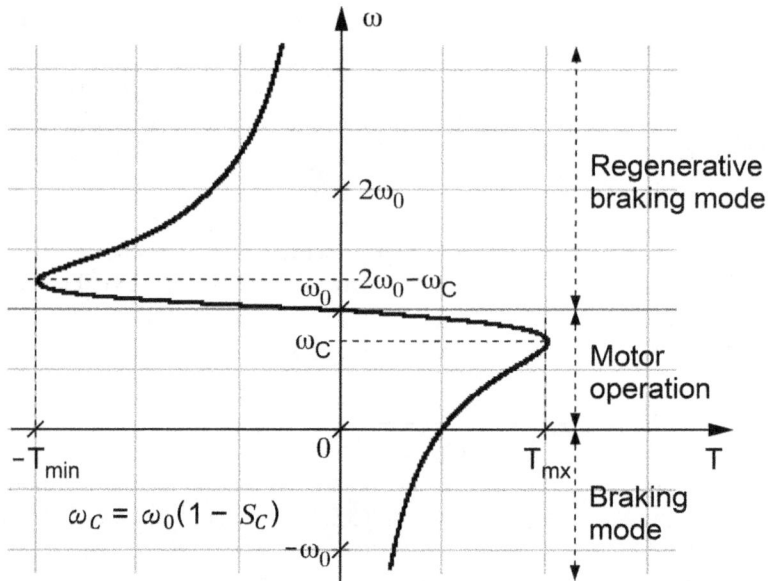

Figure 5.14. Speed versus EM torque characteristic, $\omega = f(T)$, of an induction machine.

$$\left.\begin{array}{l}
T_{mx} = \dfrac{3}{2\omega_0} \cdot \dfrac{V_{ph}^2}{R_1 + \sqrt{R_1^2 + \left(X_1 + X_2'\right)^2}} \cong \dfrac{3}{2\omega_0} \cdot \dfrac{V_{ph}^2}{X_1 + X_2'} \quad [\text{Nm}] \\[4mm]
T_{min} = \dfrac{3}{2\omega_0} \cdot \dfrac{V_{ph}^2}{R_1 - \sqrt{R_1^2 + \left(X_1 + X_2'\right)^2}} \cong \dfrac{3}{2\omega_0} \cdot \dfrac{V_{ph}^2}{-\left(X_1 + X_2'\right)} \quad [\text{Nm}]
\end{array}\right\} \quad (5.30)$$

Note that the approximation above (equation (5.30)) suggests $|T_{mx}| = |T_{min}|$.

- *The EM torque* equation (equation (5.28)) can be modified by substituting the values of the critical slip S_C (equation (5.29)) and the maximum EM torque T_{mx} (equation (5.30)):

$$\left.\begin{array}{l}
\text{Accurate value of the EM torque: } \quad T = \dfrac{2 \cdot T_{mx}\left(1 + \dfrac{R_1}{R_2'}S_C\right)}{\dfrac{S}{S_C} + \dfrac{S_C}{S} + 2\dfrac{R_1}{R_2'}S_C} \quad [\text{Nm}] \\[8mm]
\text{Approximate value: } \quad T \cong \dfrac{2\,T_{mx}}{\dfrac{S}{S_C} + \dfrac{S_C}{S}} \quad [\text{Nm}]
\end{array}\right\} \quad (5.31)$$

A comparison between the accurate and the approximate value of the EM torque equation (equation (5.31)) is presented graphically in figure 5.15.

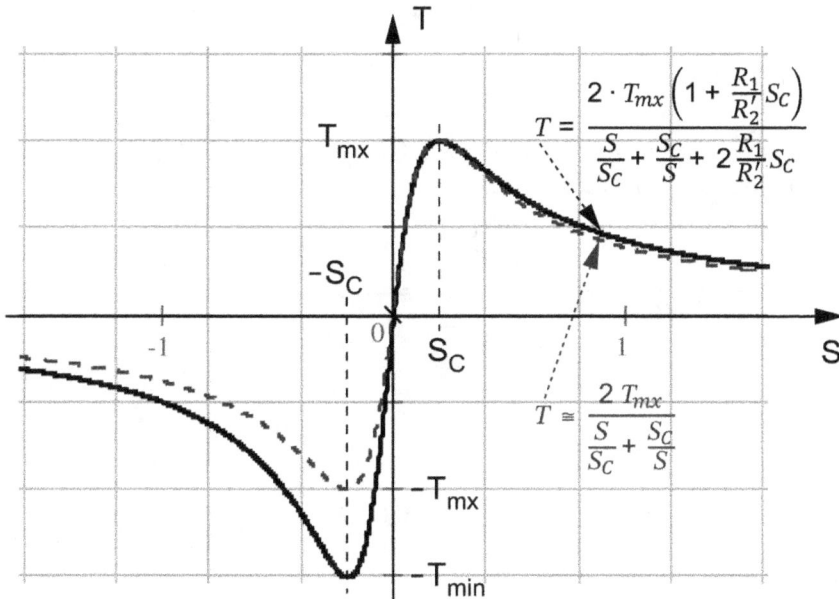

Figure 5.15. EM torque versus slip characteristics (accurate versus approximate curves).

The bold line (figure 5.15) presents the accurate torque versus slip curve of the induction machine. The dashed line presents the approximate curve. In the first quadrant, the dotted line exhibits a close match with the accurate curve, which suggests that the approximated torque equation is applicable during motoring and braking modes. In the fourth quadrant, most of the dashed blue-line curve does not concur with the bold dark-line curve. Nevertheless, the approximate equation can practically be applied in that fourth quadrant only to the operating region of the machine, which is the linear portion of the curve around the zero crossing (origin).

Example 5.1 A 3-ϕ induction motor operates at rated torque T_n and rated slip S_n. The motor has a maximum toque $2.25 \cdot T_n$, and a starting torque $1.5 \cdot T_n$. Neglect the motor losses and calculate:
  (a) the critical slip, and
  (b) the rated slip.

Solution
  (a) At start, the slip is $S = 1$. Using the approximate torque equation (equation (5.31)):

$$\frac{T_{st}}{T_{mx}} = \frac{1.5 \cdot T_n}{2.25 \cdot T_n} = \frac{2}{\dfrac{1}{S_C} + \dfrac{S_C}{1}} \quad \Longrightarrow \quad S_C^2 - 3 \cdot S_C + 1 = 0$$

That results in:

$$S_C = \begin{cases} 2.62 \text{ (unacceptable for motor operation)} \\ \\ 0.38 \end{cases}$$

  (b) At rated torque T_n, the slip is S_n. Using the approximate torque equation (equation (5.31)) once more:

$$\frac{T_n}{T_{mx}} = \frac{T_n}{2.25 \cdot T_n} = \frac{2}{\dfrac{S_n}{0.38} + \dfrac{0.38}{S_n}} \quad \Longrightarrow \quad S_n^2 - 1.72 \cdot S_n + 0.146 = 0$$

That results in:

$$S_n = \begin{cases} 1.62 \text{ (unacceptable for motor operation)} \\ \\ 0.09 \end{cases}$$

Check graphically in figure E5.1.

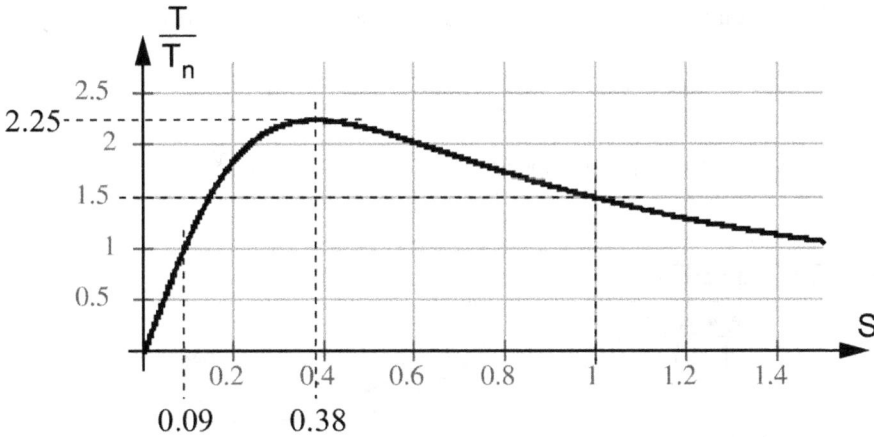

Figure E5.1. The normalized torque as a function of the slip.

5.4 Braking mode of operation

As suggested to a previous discussion (section 4.4), during braking conditions, the mechanism drives the motor shaft and the motor operates as a braker. Three types of braking modes are available [4–8]:

(a) Regenerative braking mode—when the motor operates as a generator and electric power is returned to the energy source.

(b) Dynamic braking mode—when the motor operates as a generator but the electric power is dissipated in the motor winding or in an external resistor.

(c) Plugging braking mode or 'countercurrent braking'—when the motor is operated as a generator connected in series with the utility source and the combined electric power is wasted in the motor winding or in an external resistor.

5.4.1 Regenerative braking mode

In the regenerative braking mode, the shaft rotates faster than the no-load speed of the motor. Using the approximate equation for the speed versus torque curve (equation (5.31)), one can arrive at:

$$\lambda = \frac{T_{mx}}{T} = \frac{1}{2}\left(\frac{S}{S_C} + \frac{S_C}{S}\right) \implies S_C^2 - 2 \cdot \lambda \cdot S \cdot S_C + S^2 = 0$$

$$\text{and the critical slip would be: } S_C = S(\lambda \pm \sqrt{\lambda^2 - 1})$$

$$\text{where: } S = 1 - \frac{\omega}{\omega_0}$$

$$\lambda = \frac{T_{mx}}{T} \text{ is a torque ration}$$

(5.32)

Finally, the angular speed versus the EM torque, $\omega = f(T)$, would be:

$$\omega = \omega_0\left(1 - \frac{S_C}{\lambda \pm \sqrt{\lambda^2 - 1}}\right) \quad [\text{s}^{-1}] \tag{5.33}$$

The schematic diagram and the speed versus torque characteristics are (figure 5.16).

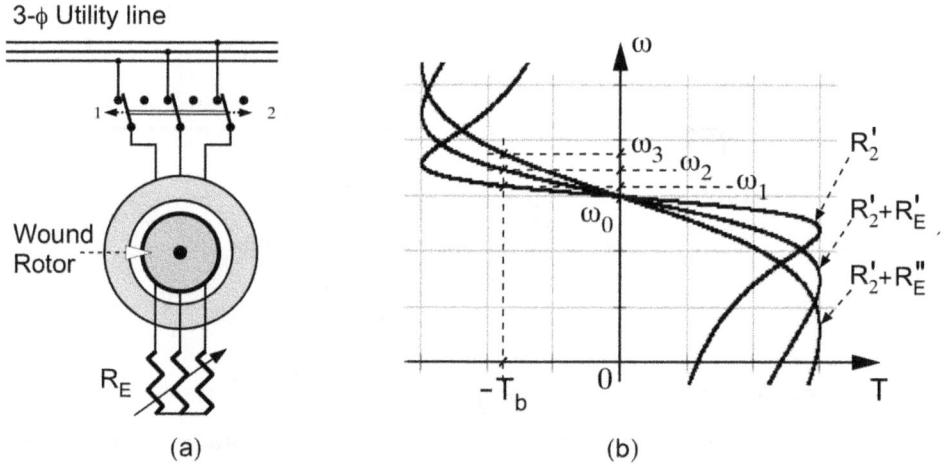

Figure 5.16. Regenerative braking of a wound rotor induction motor. (a) Circuit diagram and (b) speed versus torque curves (equation (5.33)).

A wound rotor having an external resistor(s) per phase R_E (figure 5.16(a)) allows for varying the numerical value of the critical slip:

$$S_C \cong \frac{R_2' + R_E}{X_1 + X_2'} \tag{5.34}$$

The 3-ϕ switch is in position 1 (figure 5.16(a)). When the external resistance R_E varies ($R_E'' > R_E' > 0$), at a given braking torque $-T_b$, the speed would vary too ($\omega_3 > \omega_2 > \omega_1$). The operation is depicted in the second quadrant (figure 5.16(b)).

Note: The linear curves around the synchronous speed ω_0 (figure 5.16(b)) are the actual operating region of the induction machine. That fact (linear curves) displays a close resemblance to the regenerating mode of a DC shunt connected motor (figure 4.6(b)).

Using the simplified equivalent diagram (figure 5.11), the rotor current I_2' referred to the stator is:

$$\left.\begin{array}{c} \overline{I_2'} = \dfrac{V_{ph}}{R_1 + \dfrac{R_2'}{S} + j\left(X_1 + X_2'\right)} \cong \dfrac{V_{ph}}{\dfrac{R_2'}{S} + jX} \quad [\text{A}] \\[4mm] \text{where: } X = X_1 + X_2' \end{array}\right\} \tag{5.35}$$

The simplification, $R_1 \ll \frac{R_2'}{S}$, is justified at operating speeds where the slip is only a few percentage points. The active and reactive components of the current are:

$$\overline{I_2'} \cong V_{ph} \left[\frac{R_2' \cdot S}{\left(R_2'\right)^2 + (S \cdot X)^2} - j \frac{X \cdot S^2}{\left(R_2'\right)^2 + (S \cdot X)^2} \right] \ [A] \qquad (5.36)$$

When the motor shaft speed becomes higher than its synchronous speed, $\omega > \omega_0$, the active component of the current becomes negative because $S < 0$. Suggesting that the active component reverses direction. At the same time, the sign of the reactive component remains the same (equation (5.36)).

The input current I_1 to the motor (see figure 5.11) is:

$$\overline{I_1} = \overline{I_2'} + \overline{I_m} = V_{ph} \left[\frac{R_2' \cdot S}{\left(R_2'\right)^2 + (S \cdot X)^2} - j \left(\frac{1}{X_m} + \frac{X \cdot S^2}{\left(R_2'\right)^2 + (S \cdot X)^2} \right) \right] \ [A] \quad (5.37)$$

and the apparent power to the motor is:

$$\overline{(VA)_{in}} = P_{in} + jQ_{in} = 3 \cdot V_{ph} \cdot (\overline{I_1})^* \ [VA] \qquad (5.38)$$

Conclusion: When the induction motor operates in a regenerative mode ($S < 0$), active power [W] is supplied by the motor back to the utility line, while reactive power [VAR] is supplied by the utility to the motor.

5.4.2 Dynamic braking mode

Dynamic braking is usually obtained by switching over the stator windings from a 3-ϕ voltage source to a DC voltage source (figure 5.17(a)):

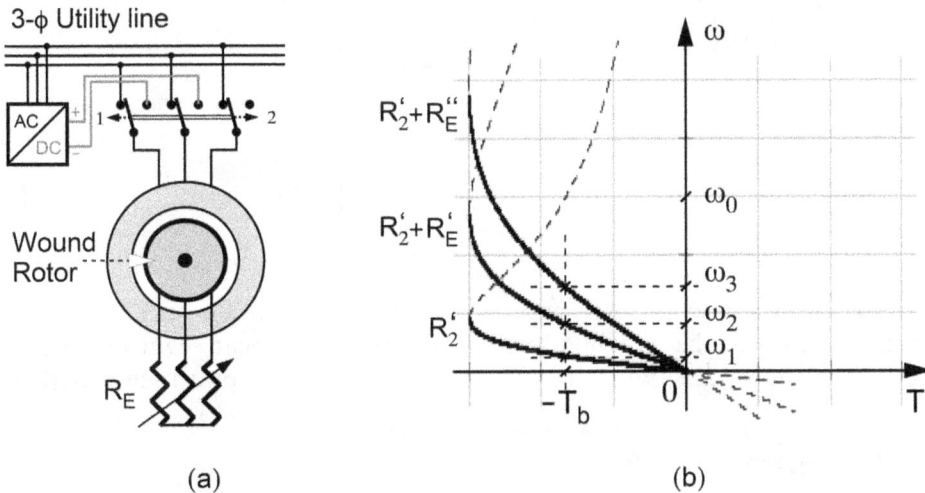

(a)   (b)

Figure 5.17. Dynamic braking of a wound rotor induction motor. (a) Circuit diagram and (b) speed versus torque curves.

When the switch is in position 1 (figure 5.17(a)), the motor operates in the first quadrant (forward motoring). In position 2, the AC/DC source (see figure 4.17) supplies DC current to the stator windings. The direct current generates a stationary magnetic flux in the motor airgap and currents are induced in the windings of the rotating rotor. The generated power dissipates in the rotor windings themselves and in external resistors (when connected). The generation action requires a negative torque from the motor shaft that opposes and reduces the motoring operation (a dynamic braking mode).

As the rotor speed decreases, the frequency of the induced currents drops down to zero frequency (standstill), so as to set up a magnetic field that is stationary with respect to the stator. In this case, the speed deference between the rotor and the stationary field is the rotor speed itself, $\Delta\omega = \omega$. By definition (equation (5.16)), the dynamic braking slip becomes:

$$S_D = \frac{\Delta\omega}{\omega_0} = \frac{\omega}{\omega_0} \tag{5.39}$$

From equation (5.16), the dynamic braking slip S_D can also be expressed as:

$$S_D = \frac{\omega}{\omega_0} = \frac{\omega_0(1 - S)}{\omega_0} = 1 - S \tag{5.40}$$

where S is the slip in the motoring operation mode.

Using equation (5.32), the critical slip becomes:

$$\left. \begin{aligned} S_C = S_D(\lambda \pm \sqrt{\lambda^2 - 1}) &= (1 - S)(\lambda \pm \sqrt{\lambda^2 - 1}) \\ \text{where: } S &= 1 - \frac{\omega}{\omega_0} \end{aligned} \right\} \tag{5.41}$$

Finally, the equation for speed versus torque during dynamic braking is:

$$\left. \begin{aligned} \omega = \omega_0 \cdot S_C &\frac{1}{\lambda \pm \sqrt{\lambda^2 - 1}} \\ \text{where: } S_C &\cong \frac{R_2' + R_E}{X_1 + X_2'} \\ \text{In this case: } \lambda &= \frac{T_{mx}}{-T} \end{aligned} \right\} \tag{5.42}$$

Note: The linear segments of the curves (second quadrant in figure 5.17(b)) are the applicable region of the dynamic braking mode. Those linear segments display a close resemblance to the dynamic braking curves of a DC motor (figure 4.7(b)).

5.4.3 Plugging braking mode

Plugging braking mode can be achieved in two ways: (1) by connecting relatively large external resistors to the wound rotor brushes to maintain constant speed. As an example, when lowering the load of a hoist mechanism while the motor is still connected so as to raise the load; and (2) by reversing of the rotating magnetic field in the airgap of the motor, so as to bring about a rapid stopping.

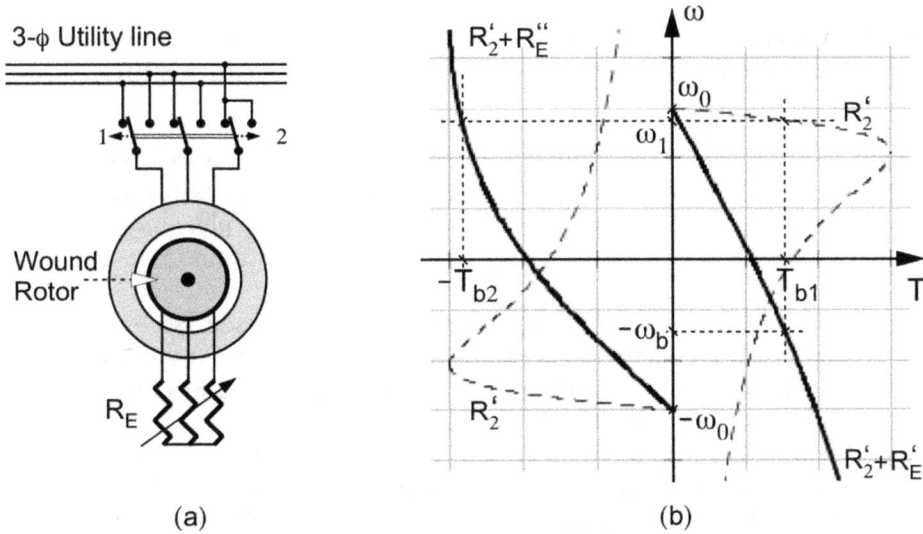

Figure 5.18. Plugging braking mode of a wound rotor induction motor. (a) Circuit diagram and (b) speed versus torque curves.

Case 1: Connecting relatively large external resistors to the wound rotor windings. The 3-ϕ switch is in position 1 (figure 5.18(a)). The added external resistors in the rotor, R'_E per-phase, reduce the EM torque T_{b1} developed by the motor below the referred load-torque at the motor shaft (fourth quadrant operation in figure 5.18(b)), and the load is lowered at a constant speed of $-\omega_b$.

Case 2: Reversal of the direction of the rotating magnetic field in the airgap of the machine. The 3-ϕ switch moves to position 2 (figure 5.18(a)), thus interchanging two phases of the stator winding while the rotor still rotates at its initial direction. Connecting external resistors, R''_E per-phase, to the rotor windings would prevent current overloading (second quadrant operation in figure 5.18(b)). This case presents an aggressive braking torque $-T_{b2}$ that can bring the drive to a complete stop (figure 5.18(b)).

The equations of the two curves in figure 5.18(b) are:

$$\left.\begin{array}{c} \text{In the 1st and 4th quadrants: } \omega = \omega_0\left(1 - \dfrac{S_C}{\lambda \pm \sqrt{\lambda^2 - 1}}\right) \ [\text{s}^{-1}] \\[3ex] \text{In the 2nd and 3rd quadrants: } \omega = -\omega_0\left(1 - \dfrac{S_C}{-\lambda \pm \sqrt{\lambda^2 - 1}}\right) \ [\text{s}^{-1}] \\[3ex] \text{where: } S_C \cong \dfrac{R'_2 + R_E}{X'_1 + X'_2} \\[2ex] \text{and: } \lambda = \dfrac{T_{mx}}{T} \end{array}\right\} \quad (5.43)$$

Note: Both characteristics (figure 5.18(b)) display a close resemblance to the plugging braking mode of a DC shunt connected motor (figure 4.8(b)).

Example 5.2 A wound rotor induction motor has the following parameters: Rated voltage $V_n = 4.16$ kV, max torque $T_{mx} = 2.25 \cdot T_n$, where T_n is the rated torque, critical slip $S_C = 0.38$, magnetizing reactance $X_m = 24$ [Ω/ph], and a combined leakage reactance $X = X_1 + X_2' = 0.53$ [Ω/ph]. Neglect the motor losses and address the following questions:

(a) The motor operates at a regenerative braking mode (second quadrant; figure 5.16) and returns energy to the supply circuit. At a braking torque $T_b = T_n$, calculate the shaft speed of ω_1 (in terms of the synchronous speed), and the active power supplied back to the utility line.

(b) Dynamic braking mode is applied to stop the motor from its rated speed (second quadrant; figure 5.17). Calculate the external resistor per phase connected to the wound rotor that would limit the starting braking torque to its rated value.

(c) The motor operates in a plugging braking mode (second quadrant; figure 5.18), where the initial speed, ω_1, is half its synchronous speed and the initial torque is its rated value. Calculate the added external resistance per phase connected in the wound rotor.

Solution

(a) *Regenerative braking mode* (second quadrant in figure 5.16(b))

Using the approximate equation (equation (5.31)), the speed versus torque curve is symmetrical with respect to the vertical axis. Therefore, for a regenerative braking torque $T_b = T_n$, the rated slip is: $-S_n$.

Using equation (5.33), the shaft speed is:

$$\frac{\omega_1}{\omega_0} = 1 - \frac{S_C}{\dfrac{T_{mx}}{-T_n} - \sqrt{\left(\dfrac{T_{mx}}{T_n}\right)^2 - 1}} = 1 - \frac{0.38}{-2.25 - \sqrt{2.25^2 - 1}} = 1.09$$

which implies that the rated slip is: $S_n = 0.09$.

Note: In the operating region of the curve (second quadrant in figure 5.16(b)), the radical is negative.

The rotor resistance is (equation (5.29)): $R_2' = X \cdot S_C = 0.53 \cdot 0.38 = 0.2$ [Ω/ph].

The slip during the braking mode is (equation (5.16)): $S = 1 - 1.09 = -0.09$.

The power at the motor terminals during the braking mode is (equation (5.38)):

$$\overline{(VA)}_{in} = 3\frac{4160}{\sqrt{3}}\left[\frac{0.2(-0.09)}{0.2^2 + (-0.09 \cdot 0.53)^2} + j\left(\frac{1}{24} + \frac{0.53 \cdot (-0.09)^2}{0.2^2 + (-0.09 \cdot 0.53)^2}\right)\right]$$

$$= -3.07 + j1.03 \text{ [kVA]}$$

and the active power supplied back to the utility line is: 3.07 kW.
(b) Dynamic braking mode (second quadrant; figure 5.17(b))
The slip in this case is:

$$S_D = \frac{\omega_n}{\omega_0} = 1.09$$

The critical speed during the braking mode is (equation (5.41)):

$$S_C = S_D(\lambda + \sqrt{\lambda^2 - 1}) = 1.09(2.25 + \sqrt{2.25^2 - 1}) = 4.65$$

Note: In the operating region of the curve (second quadrant in figure 5.17
(b)), the radical is positive.
The required external resistance is (equation (5.42)):

$$4.65 \cong \frac{0.2 + R_E}{0.53} \implies R_E \cong 2.26 \ [\Omega/\text{ph}]$$

(c) Plugging mode of operation (second quadrant; figure 5.18(b))

From equation (5.43), the critical slip is:

$$\frac{0.5 \cdot \omega_0}{-\omega_0} = \left(1 - \frac{S_C}{-(-2.25) + \sqrt{2.25^2 - 1}}\right) \implies S_C = 6.4 \ [\Omega/\text{ph}]$$

Note: In the operating region of the curve (second quadrant in figure 5.18(b)), the
radical is positive.
An optional solution: The slip at the initial plugging mode speed is:

$$S = \frac{-\omega_0 - 0.5 \cdot \omega_0}{-\omega_0} = 1.5$$

Using the approximate equation (equation (5.31)):

$$\frac{T}{T_{mx}} \cong \frac{2}{\dfrac{S}{S_C} + \dfrac{S_C}{S}} \implies \frac{1}{2.25} = \frac{2}{\dfrac{1.5}{S_C} + \dfrac{S_C}{1.5}}$$

$$S_C = \begin{cases} 6.4 \ (\text{identical to the above result}) \\ 0.352 \ (\text{unacceptable for plugging braking mode}) \end{cases}$$

Finally, the required external resistance is (equation (5.43)):

$$6.4 \cong \frac{0.2 + R_E}{0.53} \implies R_E \cong 3.19 \ [\Omega/\text{ph}]$$

5.5 Speed control of induction motors

The speed–torque formula (equation (5.33)) suggests four ways of varying the speed
of an induction motor:
(a) Varying the external resistors connected to the wound rotor of the machine.

(b) Varying the input (terminal) voltage to the motor.

(c) Varying the input frequency of the supply voltage to the motor.

(d) Changing the number of pole-pairs in the stator winding. The disadvantages of this method are: it requires multiple sets of windings in the stator, its equivalent large airgap leads to relatively high magnetizing currents, by comparison it exhibits lower efficiency, the machine is relatively large in size and weight, and its abrupt change of speed can cause an excess of mechanical wear. Consequently, a pole-changing is not a desirable technique.

5.5.1 Resistor control of wound rotor motors

Using this subject method, the synchronous speed ω_0 remains constant, but the critical slip (equation (5.34)) varies with a change in the value of the resistance R_E.

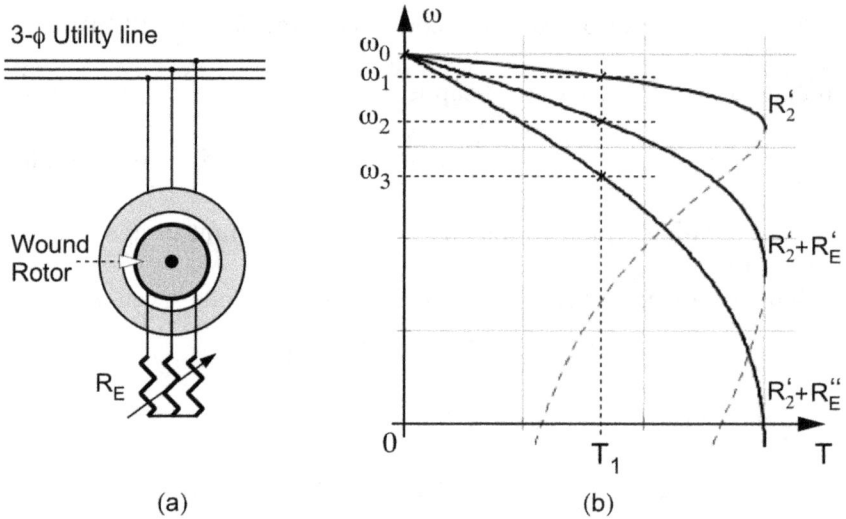

Figure 5.19. Speed control—varying external resistors to a wound rotor. (a) Circuit diagram and (b) speed versus torque curves.

In figure 5.19(a), the three external resistors vary at the same time, $R_E'' > R_E' > 0$. That would change the value of the critical slip (figure 5.19(b)). For a given EM torque T_1, the shaft speed would also change, where $\omega_3 < \omega_2 < \omega_1$.

Note that the EM torque T_{em} as well as the EM power P_{em} developed in the airgap of the motor relate directly to the square of the motor current (equation (5.25)). The copper losses P_{cu2} in the rotor also relate directly to that rotor current (equation (5.21)). Both are not a function of the shaft speed. As a result, at a given EM torque, the EM power and the copper losses in the rotor do not change with speed.

At a given EM torque T_1 (figure 5.19(b)), the motor slip at ω_2 and at ω_3 would be (equations (5.21) and (5.22)):

$$S_2 = \frac{P_{cu2}^{(2)}}{P_{em}} = \frac{3 \cdot \left(I_2'\right)^2 \left(R_2' + R_E'\right)}{P_{em}} \left.\begin{array}{c} \\ \\ \\ \\ \end{array}\right\} \quad \frac{S_2}{S_3} = \frac{R_2' + R_E'}{R_2' + R_E''} \qquad (5.44)$$

$$S_3 = \frac{P_{cu2}^{(3)}}{P_{em}} = \frac{3 \cdot \left(I_2'\right)^2 \left(R_2' + R_E''\right)}{P_{em}}$$

where

$$S_2 = 1 - \frac{\omega_2}{\omega_0} \left.\begin{array}{c} \\ \\ \end{array}\right\}$$
$$S_3 = 1 - \frac{\omega_3}{\omega_0} \qquad (5.45)$$

In practice, the external resistor(s) R_E can be controlled via an AC/DC rectifier unit [9–13].

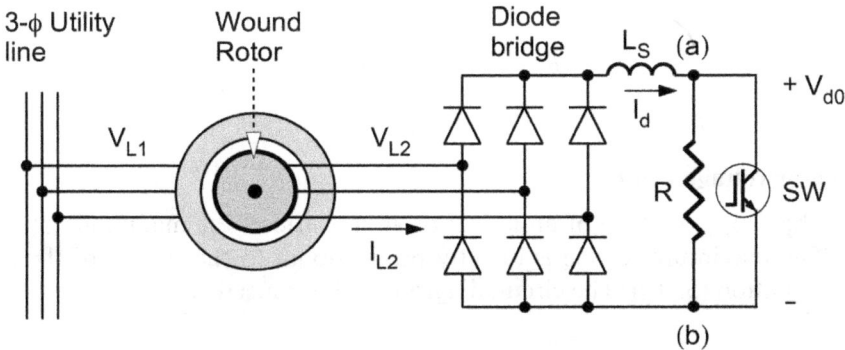

Figure 5.20. External variable DC resistor.

The rotor winding terminals are connected to a six-pulse diode rectifier (figure 5.20). The slip power is rectified and fed through a filtering inductor L_S to a pulsed resistor, which is a resistor R connected in parallel with a switch SW. An on-off cyclic operation of the switch varies the equivalent value of the resistance seen at the load terminals a and b:

$$R_{ab} = \left(1 - \frac{t_{on}}{T}\right)R = (1 - \alpha)R \quad [\Omega] \qquad (5.46)$$

where

T [s] is the switching cycle,
t_{on} [s] is the on-state duration of the switch,
$\alpha = t_{on}/T$ is the duty cycle of the switch,
R [Ω] is the actual resistor value, and
R_{ab} [Ω] is the equivalent resistance value.

To avoid fluctuation in speed, the switching cycle must be much smaller than the mechanical time constant of the motor.

Assuming constant current I_d in the inductor L_S, the rotor current I_{L2} is [9–13]:

$$\text{In a six-pulse diode rectifier:} I_{L2} = \sqrt{\frac{2}{3}} \cdot I_d \text{ [A]} \tag{5.47}$$

Neglecting losses, the conservation of power suggests that the slip power absorbed by the three external resistors R_E (figure 5.19(a)) be equal to the power absorbed by the pulsed resistor R_{eq} (equation (5.46)):

$$\left.\begin{aligned} 3 \cdot I_{L2}^2 \cdot R_E &= I_d^2 \cdot R_{ab} \\ or: \quad R_E &= \frac{1}{3}\left(\frac{I_d}{I_{L2}}\right)^2 \cdot R_{ab} \end{aligned}\right\} \tag{5.48}$$

Applying the current ratio (equation (5.47)), the equivalent value for resistor R_E would be:

$$R_E = \frac{R_{ab}}{2} \text{ [}\Omega/\text{ph]} \tag{5.49}$$

5.5.2 Stator voltage control

Varying the supply voltage of an induction motor affects the magnitude of the EM torque. The maximum torque is directly proportional to the square of the applied voltage (equation (5.30)). The circuit diagram and the speed versus torque curves are (figure 5.21).

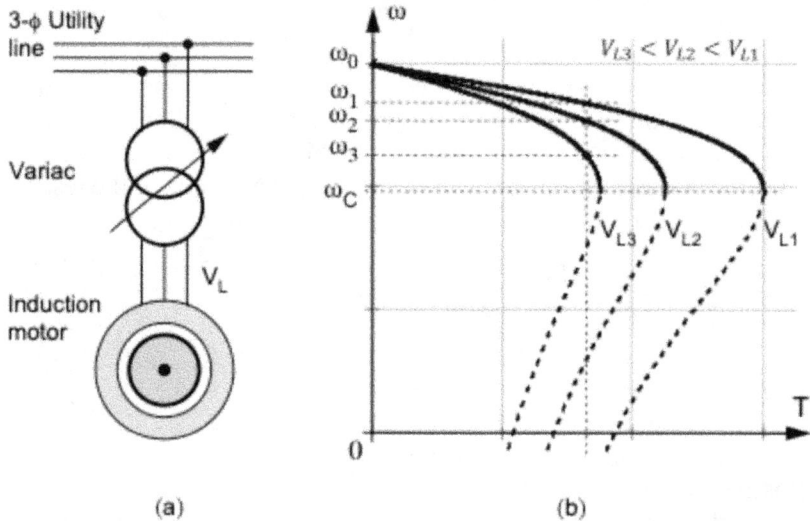

Figure 5.21. Speed control—varying the input voltage. (a) Circuit diagram and (b) speed versus torque curves.

In essence, a variable transformer (variac) can be used to adjust the input voltage of the induction motor (figure 5.21(a)). A change in the terminal voltage V_L does vary the magnitude of the maximum torque (equation (5.30)). Still, the critical slip S_C remains constant because it is independent of the terminal voltage (equation (5.29)).

As shown in figure 5.21(b), the available range of speed control is limited to $\omega_C < \omega < \omega_o$, that is between the critical speed ω_C and the synchronous speed ω_o. This mode of speed control finds use in working machines having a mechanical torque that is proportional to the square of the speed, such as with centrifugal pumps and fans.

One way to change the terminal voltage can be accomplished by inserting an inductor(s) between the voltage source and the motor terminals, as in figure 5.22.

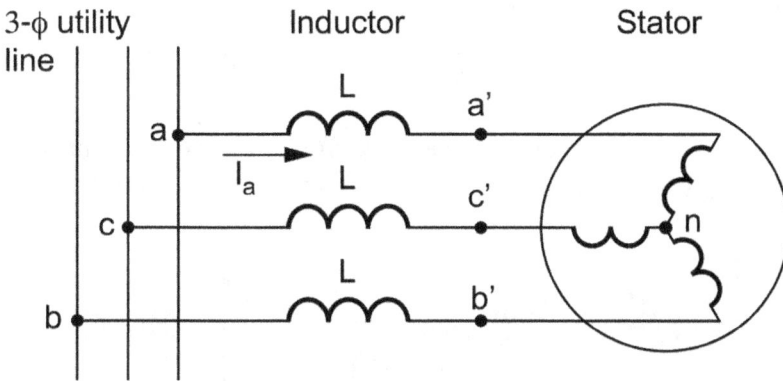

Figure 5.22. Speed control—inserting inductors to reduce the input voltage to the motor.

Neglecting the resistance of the inductor L (figure 5.22), the per-phase phasor diagram would be as in figure 5.23.

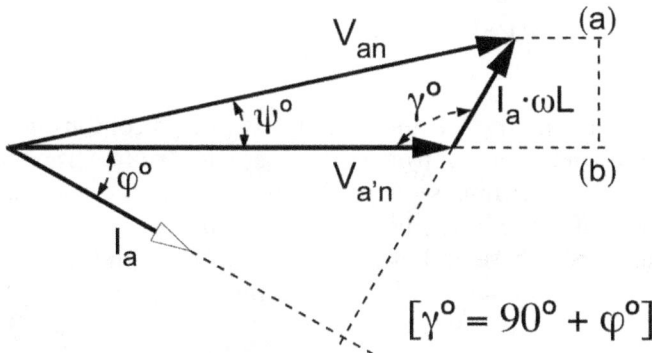

$$[\gamma° = 90° + \varphi°]$$

Figure 5.23. Per-phase phasor diagram for the circuit in figure 5.22.

In figure 5.23, the law of cosines suggests [14]:

$$\left.\begin{array}{c} V_{an}^2 = V_{a'n}^2 + (I_a \cdot \omega L)^2 - 2 \cdot V_{a'n} \cdot I_a \cdot \omega L \cdot \cos(90 + \varphi) \\ \text{or} \\ V_{a'n}^2 + 2(I_a \cdot \omega L \cdot \sin \varphi) V_{a'n} - V_{an}^2 + (I_a \cdot \omega L)^2 = 0 \end{array}\right\} \quad (5.50)$$

where

V_{an} [V] is the phase voltage of the utility line,
$V_{a'n}$ [V] is the phase voltage at the motor terminals,
ωL [Ω] is the reactance of the inductor,
I_a [A] is the line current in phase a, and
φ is the angle by which the current I_a lags the voltage $V_{a'n}$.

Using equation (5.50), the phase voltage at the motor terminals is:

$$V_{a'n} = -I_a \cdot \omega L \cdot \sin \varphi + \sqrt{(I_a \cdot \omega L \cdot \sin \varphi)^2 + V_{an}^2 - (I_a \cdot \omega L)^2} \quad (5.51)$$

Note that for a positive voltage, the negative sign of the radical is unacceptable.

The angle ψ°, between the terminal voltage $V_{a'n}$ and the utility voltage V_{an} (figure 5.24), is:

$$\left.\begin{array}{c} \overline{ab} = V_{an} \cdot \sin \psi = I_a \cdot \omega L \cdot \cos \varphi \\ \text{then} \\ \psi = \sin^{-1}\left(\dfrac{I_a \cdot \omega L}{V_{an}}\right) \cos \varphi \end{array}\right\} \quad (5.52)$$

The line current can be obtained by:

$$\left.\begin{array}{c} (I_a \cdot \omega L)^2 = V_{an}^2 + V_{a'n}^2 - 2 \cdot V_{an} \cdot V_{a'n} \cdot \cos \psi \\ \text{and the line current becomes:} \\ I_a = \dfrac{1}{\omega L}\sqrt{V_{an}^2 + V_{a'n}^2 - 2 \cdot V_{an} \cdot V_{a'n} \cdot \cos \psi} \ [\text{A}] \end{array}\right\} \quad (5.53)$$

A second way to change the terminal voltage can be accomplished by inserting a bidirectional semiconductor controller, two back-to-back thyristors, between the voltage source and the terminals of the motor (figure 5.24). The controller allows bidirectional flow of current to the motor. The inductor L_1 includes the stator leakage inductance per-phase and any additional external inductance that might be inserted in the line. The inductor resistance is omitted, and the voltage $V_{a'n}$ equals the internal voltage E across the magnetizing reactance (see figure 5.10).

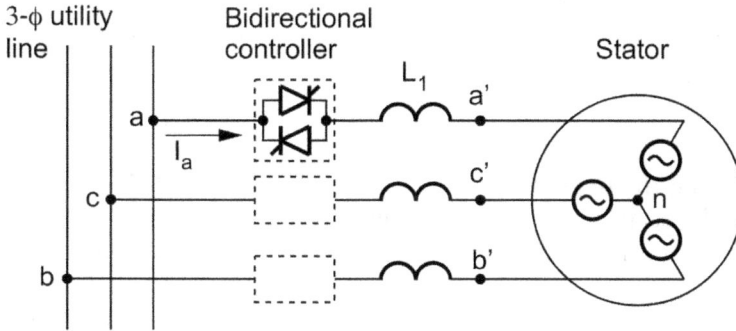

Figure 5.24. Speed control—bidirectional controller of a 3-ϕ induction motor.

Each thyristor is triggered once every cycle to obtain the required effective voltage across the motor. Both thyristors turn off naturally at current zero crossing. The instantaneous voltage and current waves are shown in figure 5.25.

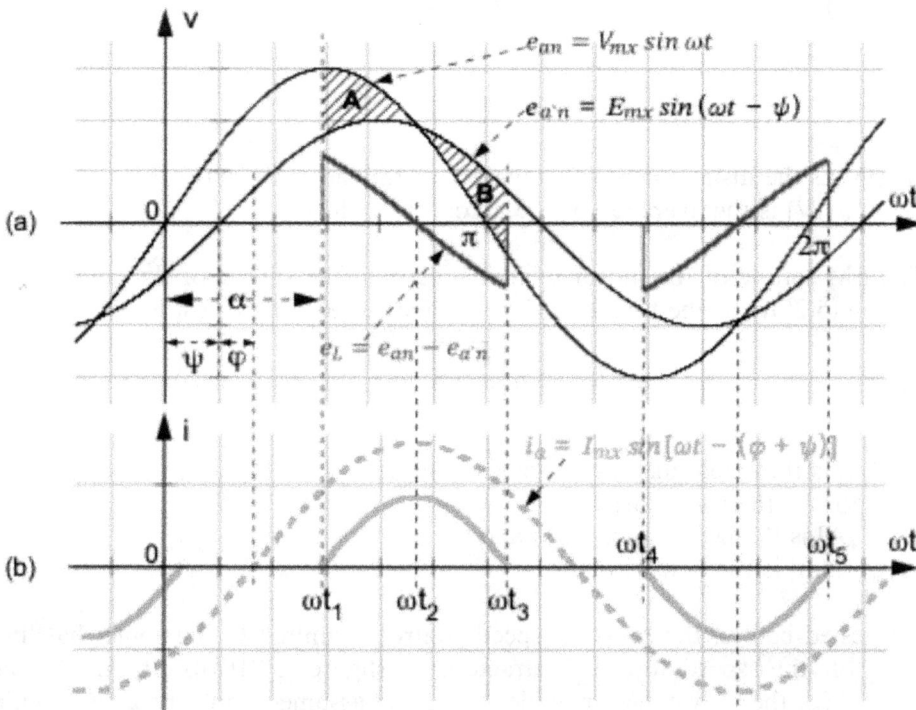

Figure 5.25. Bidirectional controller—(a) voltage waves and (b) current waves.

The internal phase voltage $e_{a'n}$ lags the utility voltage e_{an} by an angle $\psi°$ (figure 5.25(a)). The potential difference $e_L = e_{an} - e_{a'n}$ between those two voltages appears across the inductor L_1 (solid blue line).

The dashed green line in figure 5.25(b) shows a continuous line current as if the bidirectional switch was shorted out. The current peaks at the intersection between the two voltages, $e_{an}(\omega t_2) = e_{a'n}(\omega t_2)$, because the tangent of the line current is zero (equation (5.54)):

$$e_L(\omega t_2) = e_{an}(\omega t_2) - e_{a'n}(\omega t_2) = L_1 \frac{di}{dt} = 0 \tag{5.54}$$

Note that during the interval $\omega t_1 \leqslant \omega t \leqslant \omega t_3$, equation (5.53) is applicable.

When a delayed firing angle α is applied, the line current is chopped. The peak current still remains at the voltage intersection. In the first half cycle, a pulse current is shown in bold green line between ωt_1 and ωt_3 (figure 5.25(b)). In the second half cycle, the pulse current is shown between ωt_4 and ωt_5. As the current increases, the inductor L_1 is energized (area A in figure 5.25(a)), and as the current decreases, that stored energy is released (area B). At each cycle, the average voltage across the inductor must be zero (equation (5.55)):

$$\begin{aligned} V_{L,\,avg} &= \frac{1}{T} \int_{t_1}^{t_3} e_L(t) dt = \frac{1}{T}\left(\int_{t_1}^{t_2} e_L(t) \cdot dt + \int_{t_2}^{t_3} e_L(t) \cdot dt \right) \\ &= \frac{1}{T}\left(\int_{t_1}^{t_2} L_1 \frac{di}{dt} dt + \int_{t_2}^{t_3} L_1 \frac{di}{dt} dt \right) = \frac{1}{T}\left(\int_0^{I_{mx}} L_1 \cdot di + \int_{I_{mx}}^0 L_1 \cdot di \right) \\ &= \frac{1}{T}(L_1 \cdot I_{mx} - L_1 \cdot I_{mx}) = 0 \end{aligned} \right\} \tag{5.55}$$

where

e_L [V] is the instantaneous voltage across the inductor, and
$V_{L,\,avg}$ [V] is the average voltage across the inductor.

The EM torque of the motor relates directly to the square of the rotor current (equation (5.25)). As the torque varies, the line current can be estimated:

$$I_{new} \cong I_n \sqrt{\frac{T_{new}}{T_n}} \quad [A] \tag{5.56}$$

where

I_n [A] is the rated line current of the motor,
T_n [Nm] is the rated torque,
I_{new} [A] is the new line current, and
T_{new} [Nm] is the new torque.

As suggested before, the range of speed control is limited to the somewhat linear section of the speed–torque characteristic (figure 5.21(b)). In that range, $\omega_C < \omega < \omega_o$, the power factor angle φ° can be assumed constant at its nominal value. That in mind, at a new current $I_{new} \neq I_n$, the angle ψ° can be approximated (equation (5.52)). Following, the intersection time, ωt_2, can also be calculated:

$$\begin{aligned} V_{mx} \sin \omega t_2 &= E_{mx} \sin (\omega t_2 - \psi) \\ &\text{and} \\ \omega t_2 &= \tan^{-1}\left(\frac{E_{mx} \sin \psi}{E_{mx} \cos \psi - V_{mx}} \right) \end{aligned} \right\} \tag{5.57}$$

5.5.3 Stator frequency control

Varying the supply frequency to the stator changes the synchronous speed (equation (5.26)) of the induction motor. Still, in the operating region of the machine, the slop of each speed–torque curve remains constant (figure 5.26).

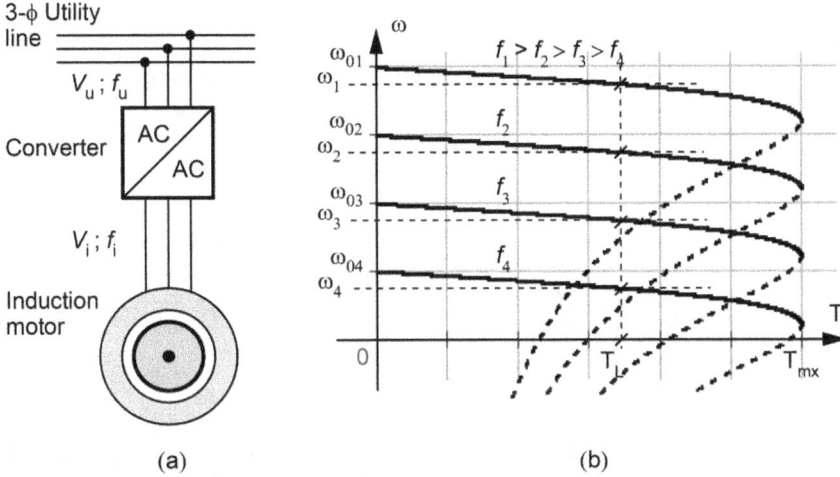

Figure 5.26. Speed control—varying the input frequency. (a) Circuit diagram and (b) speed versus torque curves.

The utility voltage V_u [V] and frequency f_u [s^{-1}] are constant. To change the supply frequency f_i to the motor terminals, an AC/AC converter is used (figure 5.26 (a)). A change in the terminal frequency would vary the synchronous speed n_0, or ω_0; that is: $\omega_{01} > \omega_{02} > \omega_{03} > \omega_{04}$ in figure 5.26(b), and, as shown in the figure, the critical slip also changes. This speed control method produces, in the operating region of the motor, a family of hard speed–torque curves parallel to the motor natural curve. This characteristic behavior is similar to the speed control of a shunt connected DC motor when varying its armature voltage (figure 4.11(c)).

The magnetic flux related directly to the phase voltage and to the reciprocal of the phase frequency (see equation (4.7)). To prevent the saturation of the iron core as the frequency is reduced, it is imperative to control the supply voltage V_i in such a manner that the magnetic flux ϕ [V s] in the airgap of the machine would remain constant [1]:

$$\phi = \frac{E}{f_i} \cong \frac{V_i}{f_i} = \text{const.} \tag{5.58}$$

Following, the approximated value of the maximum torque (equation (5.30)) relates to

$$\left. \begin{array}{l} T_{mx} \cong \dfrac{3}{2\omega_0} \cdot \dfrac{V_{ph}^2}{X_1 + X_2'} = \dfrac{3}{2\dfrac{4\pi \cdot f_i}{p}} \cdot \dfrac{V_i^2}{2\pi \cdot f_i \cdot \left(L_1 + L_2'\right)} = K\left(\dfrac{V_i}{f_i}\right)^2 \\[2em] \text{where: } K = \dfrac{3}{2\dfrac{4\pi}{p}} \cdot \dfrac{1}{2\pi\left(L_1 + L_2'\right)} \end{array} \right\} \tag{5.59}$$

Equation (5.59) suggests that keeping the airgap flux constant also maintains the maximum value of the motor torque T_{mx} (see figure 5.26(b)).

Note: The leakage reactance $X_1 + X_2'$ decreases with frequency. In the low frequencies range the approximation $R_1 \ll X_1 + X_2'$ (equation (5.29)) is no longer valid, and the voltage-drop across the stator resistance R_1 becomes significant. That causes a slight reduction in the value of the maximum torque at lower frequencies. To compensate for this change, the ratio V_i/f_i must be increased as the frequency f_i decreases.

The AC-to-AC converter itself consists of two main units. The first one is an AC/DC diode rectifier that is fed from the utility line through a power transformer and generates a constant DC voltage V_d (figure 5.27). The DC link includes a LC filter to minimize voltage fluctuations. The second unit is a pulse width modulation (PWM) inverter that generates variable AC voltage at the required frequency [9–13].

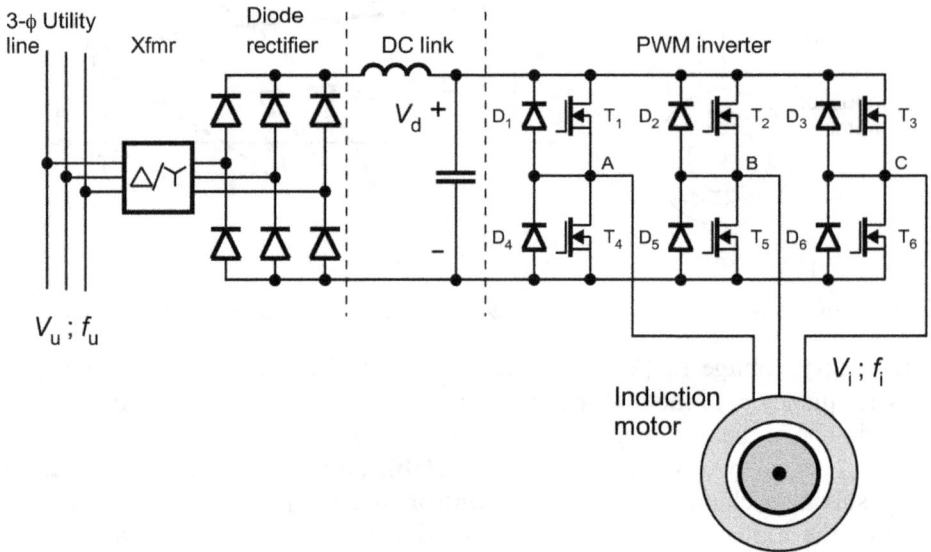

Figure 5.27. Speed control of an induction motor using an AC/AC converter schematic diagram.

The AC/DC diode rectifier (figure 5.27) has two main advantages. First, it is a reliable system that does not require any control unit. Second, the utility line sees an almost unity power factor at any load.

The DC/AC PWM inverter has the advantage of generating AC voltage with a minimal harmonic content. This unit generates a train of constant-amplitude pulses at variable pulse width. An example of an output voltage at a very low modulation frequency (pulse frequency) is presented in figure 5.28, where V_d is the DC link voltage, $1/\tau$ is the modulation frequency, and $1/T$ is the fundamental frequency.

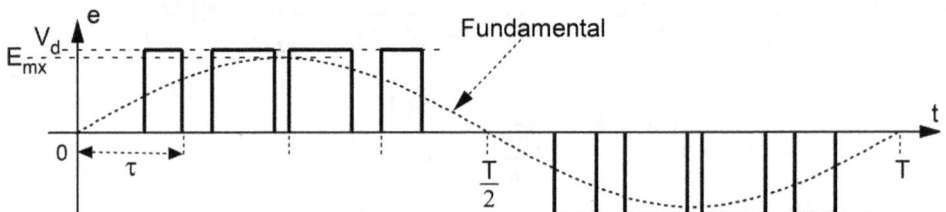

Figure 5.28. Output voltage of a PWM inverter at a low modulating frequency.

The amplitude of the individual pulses generated by the inverter is the input DC link voltage V_d (figure 5.28). The amplitude E_{mx} of the fundamental wave is controlled by varying the pulse width of each pulse. In practice, the ratio of the modulation frequency to the fundamental frequency should be maintained as high as possible to reduce the total harmonic content [9–12].

Example 5.3 A 3-ϕ wound rotor induction motor has the following nameplate parameters: $P_n = 7.5\,\text{kW}$, $V_n = 208\,\text{V}$. $I_n = 27\,\text{A}$, $n_n = 1710\,\text{rpm}$, $f_n = 60\,\text{Hz}$, $(PF)_n = 0.85$, $E_{2n} = 100\,\text{V}$, and $I_{2n} = 45\,\text{A}$. The critical slip is: $S_C = 0.25$.

The motor speed should be reduced to 1600 rpm at rated torque. Neglect the friction losses and address the following:

(a) Using external resistors in the rotor (figure 5.19)—what should be their resistance value?

(b) Using a reduced supply voltage (figure 5.21)—what should be that reduced voltage, and what external inductor(s) (figure 5.22) should be connected to the stator terminals?

(c) Using a reduced supply frequency (figure 5.26)—what should be that input (terminal) frequency, and what should be the input voltage?

Solution
Motor parameters:
The synchronous angular speed is:

$$\omega_0 = \frac{2\pi \cdot 1800}{60} = 188.5 \ [\text{s}^{-1}]$$

The nominal angular speed is:

$$\omega_n = \frac{2\pi \cdot 1710}{60} = 179.1 \ [\text{s}^{-1}]$$

The critical radial speed is:

$$\omega_C = 188.5(1 - 0.25) = 141.37 \ [\text{s}^{-1}]$$

The rated slip is (equation (5.16)):

$$S_n = \frac{188.5 - 179.1}{188.5} = 0.05$$

The maximum torque is (equation (5.31)):

$$T_n = \frac{2 \cdot T_{mx}}{\dfrac{0.05}{0.25} + \dfrac{0.25}{0.5}} \quad \Longrightarrow \quad T_{mx} = 2.6 \cdot T_n \ [\text{Nm}]$$

The rated torque is:

$$T_n = \frac{P_n}{\omega_n} = \frac{7500}{179.1} = 41.88 \ [\text{Nm}]$$

The rated efficiency is:

$$\eta_n = \frac{P_n}{P_{in}} = \frac{7500}{\sqrt{3} \cdot 208 \cdot 27 \cdot 0.85} = 0.91$$

The rotor resistance is (equation (5.24)):

$$R_2 = \frac{0.05 \cdot 100}{\sqrt{3} \cdot 45} = 0.064 \ [\Omega/\text{ph}]$$

(a) The reduced angular speed is:

$$\omega_2 = \frac{2\pi \cdot 1600}{60} = 167.55 \ [\text{s}^{-1}]$$

The new slip is:

$$S_2 = \frac{188.5 - 167.55}{188.5} = 0.111$$

The external resistor should be (equation (5.44)):

$$\frac{S_n}{S_2} = \frac{R_2}{R_2 + R_E} \implies R_E = R_2\left(\frac{S_2}{S_n} - 1\right) = 0.064\left(\frac{0.111}{0.05} - 1\right) = 0.08 \ [\Omega/\text{ph}]$$

The new critical slip is (equation (5.34)):

$$S_{C1} = S_C \cdot \frac{R_2 + R_E}{R_2} = 0.25 \cdot \frac{0.064 + 0.08}{0.064} = 0.56$$

(b) The magnitude of the maximum torque relates directly to the square of the phase voltage (equation (5.30)). Knowing that the critical slip does not vary (figure 5.21(b)), the new maximum torque would be (equation (5.31)):

$$41.88 = \frac{2 \cdot T_{mx2}}{\dfrac{0.111}{0.25} + \dfrac{0.25}{0.111}} \implies T_{mx2} = 56.46 \ [\text{Nm}]$$

The reduced voltage at the motor terminals is:

$$V_2 = V_n \cdot \sqrt{\frac{T_{mx2}}{T_{mx}}} = 208 \cdot \sqrt{\frac{56.46}{2.6 \cdot 41.88}} = 149.8 \ [\text{V}]$$

The shaft power is: $P_{out2} = T_n \cdot \omega_2 = 41.88 \cdot 167.55 = 7 \ [\text{kW}]$.

In the operating region of the motor, changes of the power factor angle and the efficiency are diminutive and practically can be neglected. Following, the line (terminal) current would be:

$$P_{out2} = P_{in2} \cdot \eta_n = \sqrt{3} \cdot V_2 \cdot I_2 \cdot (PF)_n \cdot \eta_n$$

and the line current: $I_2 = \dfrac{7000}{\sqrt{3} \cdot 149.8 \cdot 0.85 \cdot 0.91} = 34.9 \ [\text{A}]$

Note that the new line current I_2 increased well above the rated value.
Using equation (5.50):

$$V_{a'n}^2 + 2(I_a \cdot \omega L \cdot \sin \varphi) V_{a'n} - V_{an}^2 + (I_a \cdot \omega L)^2 = 0$$

$$\left(\frac{149.8}{\sqrt{3}}\right)^2 + 2\left(34.9 \cdot \frac{149.8}{\sqrt{3}} \cdot \sin(\cos^{-1} 0.85) \cdot \omega L\right) - \left(\frac{208}{\sqrt{3}}\right)^2 + 34.9^2 \cdot (\omega L)^2 = 0$$

$$(\omega L)^2 + 2.61 \cdot \omega L - 5.7 = 0$$

Finally, the reactance is:

$$\omega L = \begin{cases} -4.02 \text{ (practically, an unacceptable result)} \\ 1.42 \; [\Omega/\text{ph}] \end{cases}$$

(c) In the operating region, changes in the input frequency bring about a family of speed versus torque curves parallel to each other (figure 5.26(b)). Consequently, the reduced synchronous speed is:

$$n_{01} = 1800 - (1710 - 1600) = 1690 \; [\text{rpm}]$$

$$\text{or:} \quad \omega_{01} = \frac{2\pi \cdot 1690}{60} = 177 \; [\text{s}^{-1}]$$

The new critical slip would be (equation (5.33)):

$$\frac{167.55}{177} = 1 - \frac{S_{C1}}{2.6 \pm \sqrt{2.6^2 - 1}} \implies S_{C1} = \begin{cases} 0.011 < S_C \text{ (unacceptable)} \\ 0.27 \end{cases}$$

Using the critical slip formula (equation (5.29)), the reduced frequency is:

$$\frac{S_C}{S_{C1}} = \frac{\dfrac{R_2'}{2\pi \cdot f_n \cdot (L_1 + L_2')}}{\dfrac{R_2'}{2\pi \cdot f_1 \cdot (L_1 + L_2')}} = \frac{f_1}{f_n} \implies f_1 = 60\frac{0.25}{0.27} = 55.5 \; [\text{Hz}]$$

And the reduced input voltage would be:

$$V_2 = V_n \frac{f_1}{f_n} = 208\frac{55.5}{60} = 192.4 \; [\text{V}]$$

A *graphic presentation* of the above three cases is shown in figure E5.3.

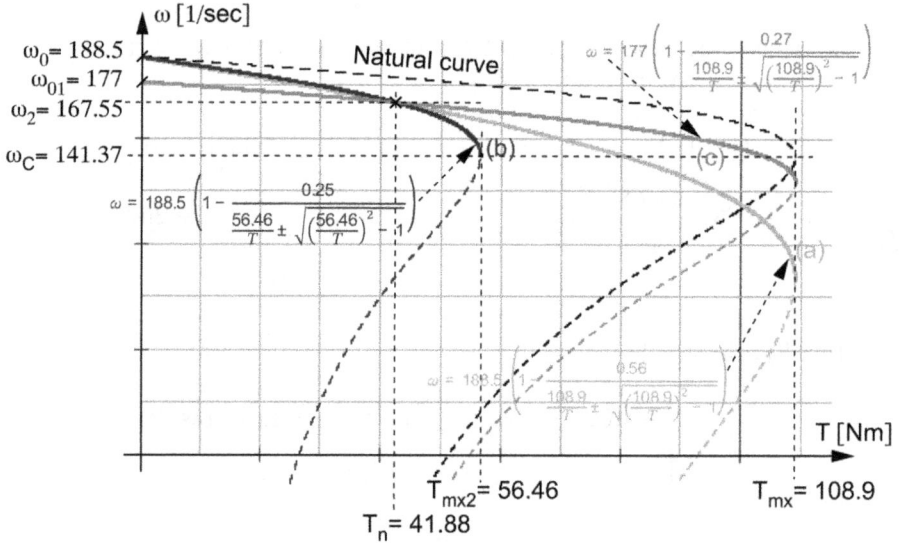

Figure E5.3. Three speed control cases of induction motor. (a) Varying external resistors, (b) varying input voltage, and (c) varying input frequency.

5.6 Problems

1. A 3-ϕ wound rotor induction motor: 460 V, 1710 rpm, 60 Hz, four poles, has the following equivalent circuit parameters: $R_1 = 0.5\,[\Omega/\text{ph}]$, $R_2' = 0.6\,[\Omega/\text{ph}]$, $X_m = 32\,[\Omega/\text{ph}]$, $X_1 = X_2' = 1.3\,[\Omega/\text{ph}]$, and a stator-to-rotor voltage ratio of 2.

   The motor operates in a regenerative braking mode (second quadrant in figure 5.16), where the braking torque equals the motor rated torque. The rotor slip rings are connected to external resistors $R_E = 0.25\,[\Omega/\text{ph}]$.

   Neglect motor friction and core losses and calculate:
   (a) the speed at which the motor operates, and
   (b) the power at the motor terminals during the braking mode.

2. A 3-ϕ squirrel-cage induction motor: 460 V, 1710 rpm, 60 Hz, six poles, has the following equivalent circuit parameters: $R_1 = 0.5\,[\Omega/\text{ph}]$, $R_2' = 0.6\,[\Omega/\text{ph}]$, $X_m = 32\,[\Omega/\text{ph}]$, $X_1 = X_2' = 1\,[\Omega/\text{ph}]$. The motor operates in a regenerative braking mode (second quadrant), where the braking torque is $T_B = -160$ Nm.

   Neglect motor friction and core losses and calculate:
   (a) the speed at which the motor operates, and
   (b) the power at the motor terminals during the braking mode.

3. A 3-ϕ wound rotor induction motor has the following parameters: $\omega_0 = 188.5\,[\text{s}^{-1}]$, $S_n = 0.05$, $S_C = 0.194$, $T_{mx} = 219.2$ Nm, and the resistance of the rotor winding is $R_2 = 0.15\,[\Omega/\text{ph}]$.

   The motor drives a mechanism at a shaft speed of $n_1 = 1710$ rpm.

A plugging braking mode is applied by abruptly interchanging two phases of the stator winding to bring the working machine to a standstill (figure 5.18). The initial braking torque equals to the rated torque of the motor.

Neglect motor friction and core losses and calculate the value of the external resistor(s) per phase that must be connected to the slip rings of the wound rotor.

4. A 3-ϕ wound rotor induction motor: 460 V, 1710 rpm, 60 Hz, four poles, has the following equivalent circuit parameters: $R_1 = 0.5$ [Ω/ph], $R_2' = 0.6$ [Ω/ph], $X_1 = X_2' = 1.3$ [Ω/ph], and a stator-to-rotor voltage ratio of 1.

The motor operates in a regenerative braking mode (second quadrant), at a speed of $n_1 = 2040$ rpm, and the braking torque equals the motor rated torque. The rotor slip rings are connected to external resistors $R_E = 1$ [Ω/ph].

Neglect friction and core losses, and *calculate* the torque at the motor shaft.

5. A 3-ϕ wound-rotor induction motor has the following parameters: $n_n = 1746$ rpm, $p = 4$ poles, $f_1 = 60$ Hz, $E_{2n} = 187$ V, and $I_{2n} = 27$ A. The equivalent leakage reactance is: $X_1 + X_2' = 0.833$ [Ω/ph]. The motor operates at its rated torque T_n and its rated speed n_n. A dynamic braking mode is applied (see figure 5.17), where the initial braking toque is: $T_b = -1.5 \cdot T_n$.

Assume that the stator-to-rotor voltage ratio is 1, and *calculate* the value of the external resistor per-phase that needs to be connected to the rotor to carry out that braking task.

6. A 3-ϕ wound rotor induction motor lifts a load at constant speed with the aid of a winding drum (figure P5.6). The motor synchronous speed is $n_0 = 900$ rpm, its combined leakage reactance is $X = X_1 + X_2' = 0.53$ [Ω/ph], and its maximum torque is twice its rated torque. The motor rotates at its rated speed while the load is lifted at a linear velocity of $v = 4$ [m s^{-1}]. The referred load-torque at the motor shaft equals the rated motor torque. The gear ratio is $i = 8$. The winding drum diameter is $D = 0.7$ m.

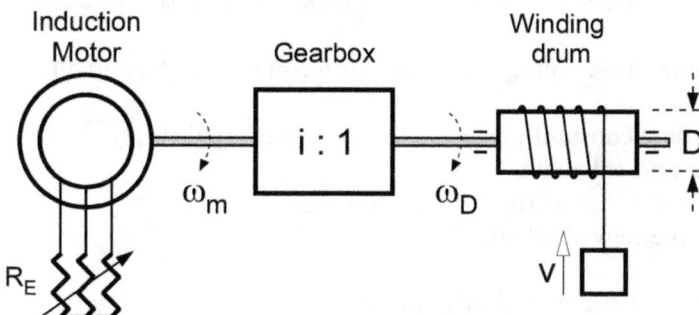

Figure P5.6. Induction motor drives a winding drum.

Maintaining rated torque at the motor shaft, the elevation speed has to be reduced to $v_2 = 3.5$ [m s^{-1}] with the aid of external resistors connected to the rotor slip rings.

Calculate the value of the external resistor(s) per-phase R'_E referred to the stator.

7. A 3-ϕ wound rotor induction motor operates at rated torque and rated speed. It has the following parameters: 5 kW, 208 V, 1710 rpm, 60 Hz, and 4 poles. The motor total equivalent leakage reactance is $X_1 + X'_2 = 0.53$ [Ω/ph], and its maximum torque is twice its rated torque.

   To reduce the motor speed at a reduced torque, three external resistors were connected to the rotor slip rings varying the critical slip to $S_{C1} = 0.24$. Calculate the value of the external resistor(s) per-phase transferred to the stator.

8. A 3-ϕ squirrel-cage induction motor drives a working machine. The motor parameters are: $P_n = 3.4$ kW, $V_n = 208$ V, $n_n = 1746$ rpm, $I_n = 12$ A, $(PF)_n = 0.877$, 60 Hz, four poles, and maximum torque equals twice its rated torque.

   To lower the speed, three identical inductors were connected in series with the motor terminals (figure P5.8).

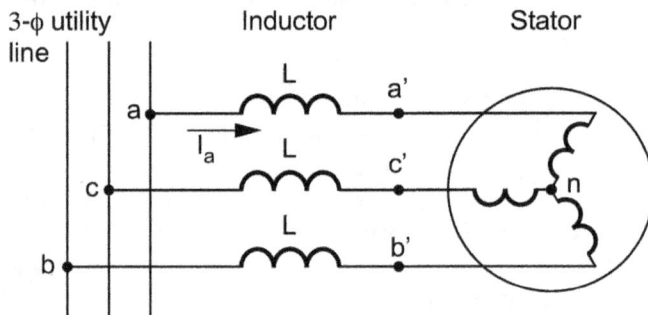

Figure P5.8. Speed control by reducing the terminal voltage.

The reduced voltage at the motor terminals (a', b', and c') is 160 V line-to-line.

Assume constant power factor, constant rated current, and constant rated torque and calculate:
   (a) the value of the reactance(s) ωL, and
   (b) the new shaft speed.

9. An electric drive system employs a semiconductor (static) converter that controls the voltage and frequency of a 3-ϕ squirrel-cage induction motor

(figure P5.9). The motor has the following parameters: 208 V, 5 kW, 1710 rpm, 60 Hz, four poles, and a maximum torque equals to twice its rated torque. Given rated torque at the motor shaft, the shaft speed n is decreased by reducing the input frequency at the motor terminals to 45 Hz.

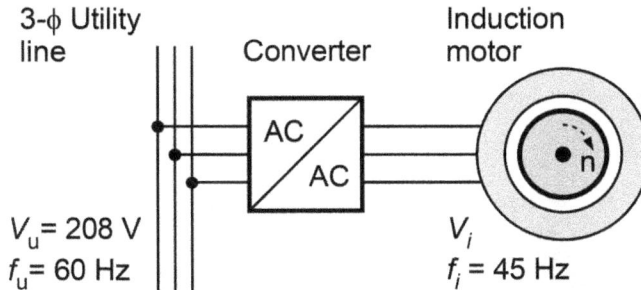

Figure P5.9. Speed control by changing the input frequency.

Neglect the friction losses and, at $f_i = 45$ Hz input frequency, calculate:

(a) the critical speed of the motor,
(b) the actual shaft speed,
(c) the voltage V_i at the motor terminals, and
(d) the power at the motor shaft.

References

[1] Fitzgerald A E, Kingsley C and Umans S 2002 *Electric Machinery* 6th edn (New York: McGraw-Hill)
[2] Kostenko M and Piotrovsky L 1974 *Electrical Machines* **2 volumes** 3rd edn (Moscow: MIR Publishers)
[3] Wildi T 2006 *Electrical Machines, Drive and Power Systems* 6th edn (Englewood Cliffs, NJ: Prentice-Hall)
[4] Mohan N 2001 *Electric Drives, An Integrative Approach* (MNPERE Publisher)
[5] Dubey G K 2001 *Fundamentals of Electrical Drives* (Alpha Science International Ltd)
[6] Nasar S A and Unnewehr L E 1983 *Electromechanical and Electric Machines* (New York: Wiley)
[7] Meyers R A (ed) 2002 *Encyclopedia of Physical Science and Technology* **vol 5** 3rd edn (New York: Academic)
[8] Chilikin M 1976 *Electric Drive* (Moscow: MIR Publishers)
[9] Erickson R W and Maksimović D 2020 *Fundamentals of Power Electronics* 3rd edn (Berlin: Springer)
[10] Schaefer J 1965 *Rectifier Circuits: Theory and Design* (New York: Wiley)
[11] Rashid M H 2004 *Power Electronics: Circuits, Devices, and Applications* 3rd edn (Englewood Cliffs, NJ: Pearson Prentice-Hall)

[12] Csaki F, Ganszky K, Ipsits I and Marti S 1975 *Power Electronics* (Gudapest: Akademiai Kiado)

[13] Zabar Z 2022 *Fundamentals of Distributed Generation Systems* (New York: AIP Publishing)

[14] Zwillinger D 2012 *Standard Mathematical Tables and Formulae* 32nd edn (Boca Raton, FL: CRC Press)

Chapter 6

The synchronous motor as a brushless DC motor

The basic torque–speed characteristics of the DC motor having an external magnetic field can be achieved by a closed-loop control of a synchronous motor where the motor is fed by a variable-frequency converter. The main advantage of that concept is the elimination of the mechanical brush-commutator unit employed by DC motors. This chapter addresses briefly the principle of operation of the synchronous motor, and the use of closed-loop variable-frequency converters.

6.1 Principle of operation

The synchronous motor has two main parts: a stationary part called the stator, and a rotating part called the rotor. *The stationary part* is similar to that of the induction motor. Basically, it has three independent coils that are spaced evenly around the stator at 120° apart from each other (figure 6.1(a)). Figure 6.1(b) shows a wye connection of those three stator windings [1, 2].

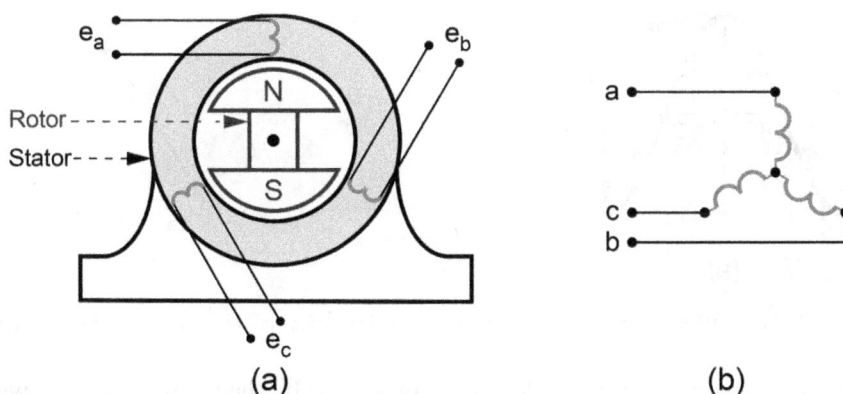

Figure 6.1. Two-poles, three-phase synchronous motor having a salient rotor. (a) Basic structure, and (b) Stator winding connection.

doi:10.1088/978-0-7503-6104-0ch6
6-1

When fed by a 3-ϕ system of voltages, e_a, e_b, and e_c (figure 6.1), those three coils generate a rotating magnetic field in the airgap, the same way as with the induction motor. As presented in chapter 5, section 5.1.2, the speed of that rotational magnetic field is:

$$\left. \begin{array}{c} n = \dfrac{120 \cdot f}{p} \ [\text{rpm}] \\ \text{or} \\ \omega = \dfrac{4\pi \cdot f}{p} \ [\text{s}^{-1}] \end{array} \right\} \tag{6.1}$$

where

n [rpm] is the speed of the rotational magnetic field in the airgap of the motor,
$f\,[\text{s}^{-1}]$ is the frequency of the input current,
p is the number of magnetic poles, and
$\omega[\text{s}^{-1}]$ is the angular speed of the magnetic field.

The rotating part generates its own magnetic field by either using a DC excitation winding (field winding) wound around the center core of the rotor, where the DC current is fed through stationary brushes and slip-rings attached to the rotating shaft; or by employing a permanent magnet. Figure 6.1(a) shows a salient structure of the rotor.

At rated speed, the rotor is locked onto the rotating magnetic field and revolves along with it. To put it another way, the rotation of the rotor is synchronized with the revolving field.

6.1.1 Salient magnetic pole action

Considering the two-poles salient rotor machine in figure 6.1(a). A linear presentation of the magnetic flux distribution in the airgap along the periphery of the machine is presented below (figure 6.2(a)):

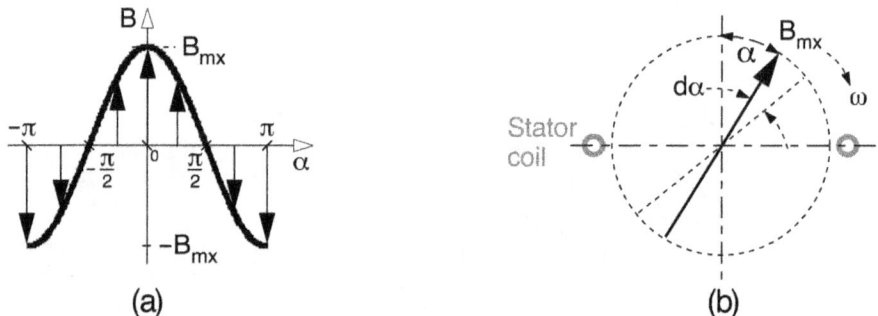

Figure 6.2. Magnetic flux density distribution with a salient rotor. (a) Distribution and (b) dynamic.

In practice, the magnetic flux density B produced by the rotor excitation winding along the periphery of the airgap is sinusoidal. That is:

$$B = B_{mx} \cos \alpha \; [\text{Vs m}^{-1}] \tag{6.2}$$

where
   B_{mx} is the maximum magnetic flux density and
   α is an angle along the periphery of the airgap.

As the rotor rotates at an angular velocity ω, an infinitesimal increment $d\varphi$ in the magnetic flux across an element of the stator surface would be (figure 6.2(b)):

$$d\varphi = B \cdot dA = B \cdot \left(\frac{D}{2} \cdot L \cdot d\alpha \right) = \frac{D}{2} \cdot L \cdot B_{mx} \cos \alpha \cdot d\alpha \; [\text{V s}] \tag{6.3}$$

where:
   D [m] is the inner diameter of a stator coil and
   L [m] is the axial length of a stator coil.

The total magnetic flux links with the stator coil would be:

$$\phi_{mx} = \int_{-\frac{\pi}{2}}^{\frac{\pi}{2}} \frac{D}{2} \cdot L \cdot B_{mx} \cdot \cos \alpha \cdot d\alpha = B_{mx} \cdot D \cdot L = B_{mx} \cdot A \; [\text{V s}] \tag{6.4}$$

where
   ϕ_{mx} [V s] is the maximum magnetic flux links with the stator coil, and
   A [m^2] is the area of the coil.

Using equations (6.2) and (6.4), the magnetic flux that links with the stator coil is:

$$\varphi = B \cdot A = \phi_{mx} \cdot \cos \alpha \; [\text{V s}] \tag{6.5}$$

Applying Faraday's law, the induced voltage $e(t)$ in the coil would be:

$$e(t) = -N \frac{d\varphi}{dt} = -N \frac{d\varphi}{d\alpha} \cdot \frac{d\alpha}{dt} = N \cdot \phi_{mx} \cdot \sin \alpha \cdot \frac{d\alpha}{dt} \; [\text{V}] \tag{6.6}$$

where N is the number of turns in the stator coil.
   The angular speed ω is (figure 6.2(b)):

$$\omega = \frac{d\alpha}{dt} \; [\text{s}^{-1}] \tag{6.7}$$

Using equation (6.6), the induce voltage in the stator coil would be:

$$\left. \begin{array}{c} e(t) = E_{mx} \cdot \sin \omega t \; [\text{V}] \\ \text{where} \\ E_{mx} = N \cdot \phi_{mx} \cdot \omega \; [\text{V}] \text{ is the maximum induced voltage} \\ \text{Also, } E = \dfrac{E_{mx}}{\sqrt{2}} \; [\text{V}] \text{ is the effective value of the induced voltage} \end{array} \right\} \tag{6.8}$$

Given a constant magnetic flux, $\phi_{mx} =$ const., such as with a permanent magnet, the induced voltage E in the stator coil would be a direct function of the rotor speed ω.

6.1.2 The equivalent diagram

Consider a 3-ϕ synchronous motor. The per phase equivalent circuit where the current lags the input voltage is [3–5].

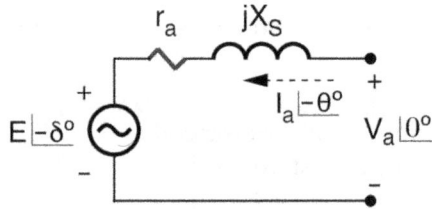

Figure 6.3. Per phase equivalent circuit of a synchronous motor at a lagging power factor.

In figure 6.3:
  E [V/ph] is the effective value of the induced voltage (equation (6.8)),
  r_a [Ω/ph] is the ohmic resistance,
  X_S [Ω/ph] is the synchronous reactance,
  I_a [A] is the input current,
  V_a [V/ph] is the input (terminal) voltage,
  $\theta°$ is the angle between the input voltage and input current, and
  $\delta°$ is the angle between the input voltage and the induced voltage.

The synchronous reactance is much larger than the ohmic resistance, $X_S \gg r_a$. In practice, a simplified equivalent circuit (figure 6.4(a)) is used.

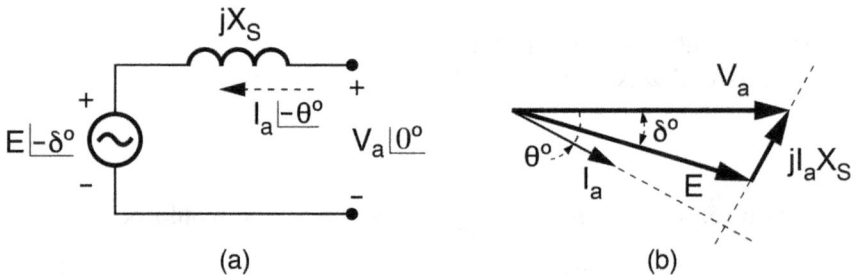

Figure 6.4. Simplified equivalent circuit of synchronous motor at a lagging power factor. (a) Per phase simplified circuit and (b) phasor diagram.

Applying Kirchhoff voltage law (KVL), the phasor diagram (figure 6.4(b)) suggests:

$$\overline{V_a} = \overline{E} + j\ \overline{I_a}X_S \text{ [V]} \tag{6.9}$$

6.1.3 The EM power and torque

- EM power:

    Using the simplified equivalent diagram (figure 6.4(a)), the input power per phase can be expressed as:

$$P = V_a \cdot I_a \cdot \cos \theta° \ [\text{W}] \qquad (6.10)$$

Neglecting the friction losses, that power (equation (6.10)) is the EM power of the motor.

Using the phasor diagram (figure 6.4(b)), one can deduce:

$$E \cdot \sin \delta = I_a \cdot X_S \cdot \cos \theta° \implies \cos \theta° = \frac{E \cdot \sin \delta°}{I_a \cdot X_S} \qquad (6.11)$$

Substitute $\cos \theta$ (equation (6.11)) into equation (6.10), the EM power per phase becomes:

$$P = \frac{V_a \cdot E}{X_S} \sin \delta° \ [W] \qquad (6.12)$$

- EM torque:

    Using the phasor diagram of the simplified equivalent circuit (figure 6.4 (b)), the EM power can also be expressed as:

$$P = E \cdot I_a \cdot \cos (\theta° - \delta°) \ [\text{W}] \qquad (6.13)$$

and the EM torque per phase would be:

$$T = \frac{P}{\omega} = \frac{E}{\omega} \cdot I_a \cdot \cos (\theta° - \delta°) \ [\text{Nm}] \qquad (6.14)$$

where $\omega \ [\text{s}^{-1}]$ is the radial speed of the rotor.

Substituting the effective value of the induced voltage E (equation (6.8)) into equation (6.14), the torque per phase becomes:

$$T \cong K_T \cdot \phi_{mx} \cdot I_a \cdot \cos (\theta° - \delta°) \ [\text{Nm}] \qquad (6.15)$$

where $K_T = N/\sqrt{2}$

Note: With a permanent magnet rotor, the magnetic flux is constant, $\phi_{mx} = \text{const}$.

6.2 Speed control

At steady state operation, the rotor of the synchronous motor rotates in synchronism with the revolving magnetic field. The synchronous speed relates directly to the frequency of the supply current (equation (6.1)). Inevitably, a change in the supply frequency would bring about a change in the synchronous speed.

The change in the supplied frequency must also coincide with a change in the supply voltage. As Faraday's law suggests (equation (6.8); also addressed in section 5.5.3), it is imperative to control the supply voltage in such a manner that the magnetic flux in the airgap of the machine would remain constant [1].

In practice, a pulsed width modulation (PWM) converter is used to vary both, the supply frequency as well as the supply voltage, to maintain constant magnetic flux in the airgap of the motor [3–6]. Two options for the variable-frequency drive are feasible: open and closed-loop control.

6.2.1 Open-loop system

In an open-loop system, the AC/AC converter controls the motor speed without information about the actual shaft speed (figure 6.5).

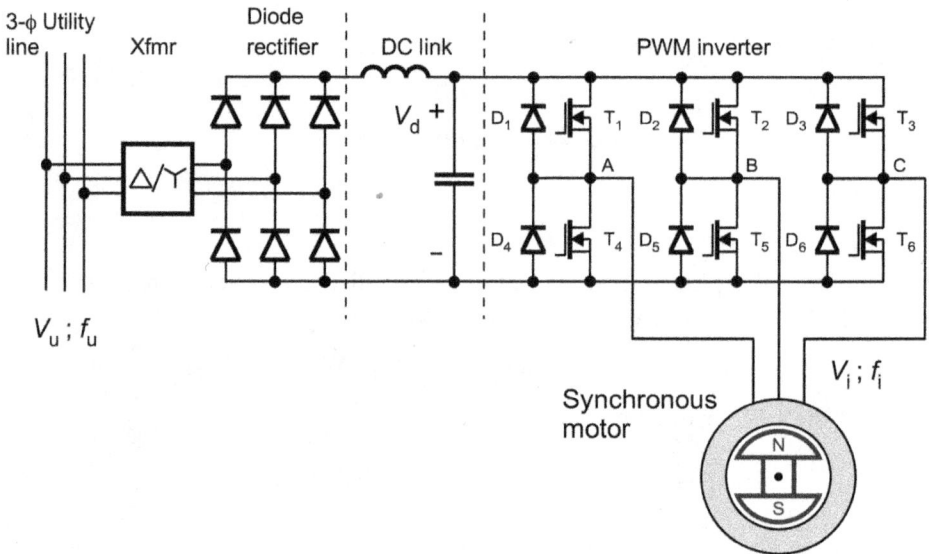

Figure 6.5. Speed control of a synchronous motor using an AC/AC converter schematic diagram.

The schematic diagram (figure 6.5) is similar to the one suggested for controlling the speed of the induction motor (figure 5.27). In both cases, a change in the supply frequency f_i brings about a change in the synchronous speed of the motor, $\omega_{01} > \omega_{02} > \omega_{03} > \omega_{04}$ in figure 6.6. To maintain constant magnetic flux ϕ in the airgap of the machine, the input voltage must also change (see equation (5.58)). That is:

$$\phi = \frac{V_i}{f_i} = \text{const.} \tag{6.16}$$

The main difference between the two is that with the induction motor, the rotor speed differs from its synchronous speed. With the synchronous motor, the rotor speed is synchronized with the rotating magnetic field.

Figure 6.6 shows four curves at different input frequency, $f_1 > f_2 > f_3 > f_4$. Compared with the speed versus torque curves of the induction motor (figure 5.26), the speed of the synchronous motor for a given frequency remains constant as the load torque increases.

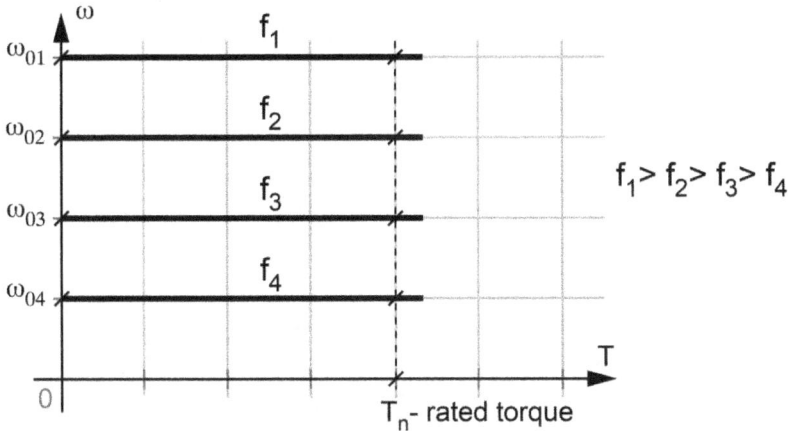

Figure 6.6. Open-loop speed control of a synchronous motor—speed versus torque.

Note: Although the flat speed curves are desirable (figure 6.6), the open-loop speed control concept of a synchronous motor has three major limitations: the motor requires a starting mechanism, it loses synchronization at overloads, and it might lose synchronization at an abrupt change in the supply frequency. Those three limitations make the open-loop speed control concept an impractical option.

6.2.2 Closed loop control

In a closed-loop control, the PWM inverter uses information about the rotor position, or the shaft speed, to adjust the input frequency and voltage that are supplied to the motor terminals [3, 4].

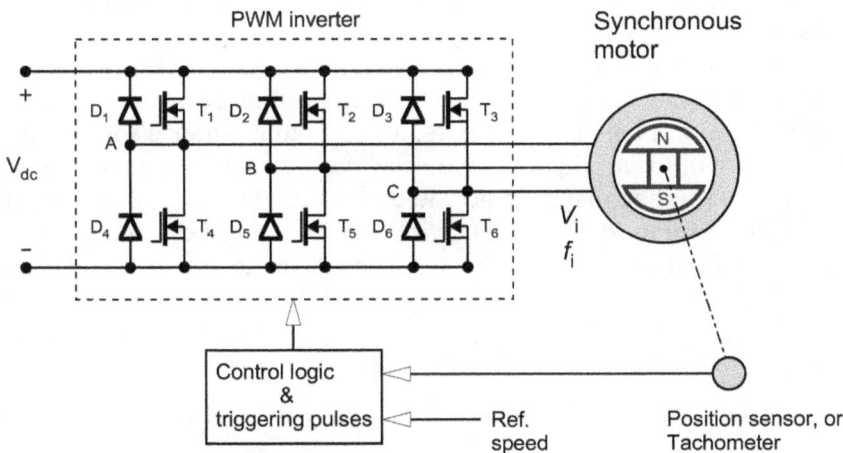

Figure 6.7. Closed loop control of a synchronous motor—a BLDC motor.

The control logic unit (figure 6.7) receives information from a rotor-position sensor, or from a tachometer, and compares it with a reference speed-signal. The difference generates pulses that trigger the PWM inverter switches at a frequency correlated with the motor speed. That way, the rotating magnetic field in the airgap of the machine would remain synchronized with the rotor speed. The operation of that self-synchronizing system resembles the operation of a DC motor having an external excitation (see figure 4.22) suggesting the name brushless DC motor, BLDC, motor [5–7].

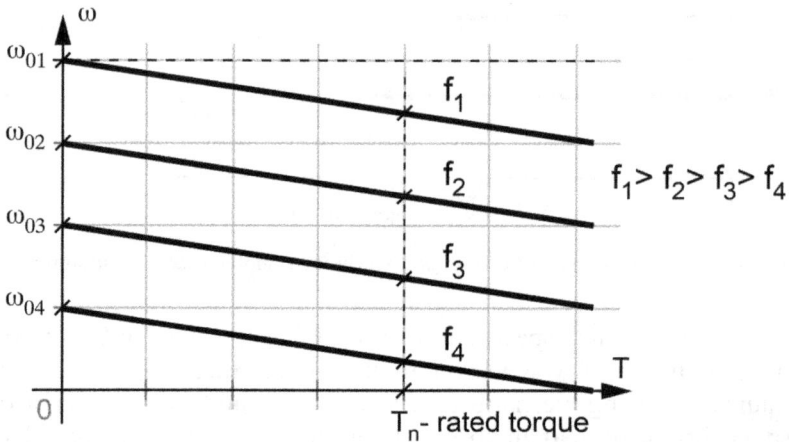

Figure 6.8. Closed loop control of a synchronous motor—speed versus torque curves.

A change in the supply frequency f_i would bring about a change in the synchronous speed of the motor, $\omega_{01} > \omega_{02} > \omega_{03} > \omega_{04}$ (figure 6.8). An increase in the load torque might reduce the shaft speed, but the inverter output frequency would also be reduced because it is based on the that speed. Consequently, the rotor (a permanent magnet or a field winding) would remain synchronized with the rotating magnetic field. As suggested above (equation (6.16)), the supplied voltage would also vary by the PWM inverter.

Modern brushless DC motors operate on the same principle as the closed-loop synchronous motor system [8–10]. That is, using position sensors, coils around the periphery of the motor are energized in sequence so to generate a rotating magnetic field in the airgap of the machine. The rotating part, usually a permanent magnet, is locked on to the rotating magnetic field and go around with it, suggesting that the rotation of the rotor is synchronized with the revolving field. The speed versus torque characteristics of the BLDC motors are similar to the curves of the closed-loop control of the synchronous motor (figure 6.8).

References

[1] Fitzgerald A E, Kingsley C and Umans S 2002 *Electric Machinery* 6th edn (New York: McGraw-Hill)

[2] Kostenko M and Piotrovsky L 1974 *Electrical Machines* **2 volumes** 3rd edn (Moscow: MIR Publishers)

[3] Novotny D W and Lipo T A 1996 *Vector Control and Dynamics of AC Drives* (Oxford: Oxford University Press)

[4] Leonhard W L 2001 *Control of Electrical Drives* 3rd edn (Berlin: Springer)

[5] Bose B K 2002 *Modern Power Electronics and AC Drives* (Englewood Cliffs, NJ: Prentice-Hall)

[6] Krause P, Wasynczuk O, Sudhoff S and Pekarek S 2013 *Analysis of Electric Machinery and Drive Systems* (New York: Wiley)

[7] Giri F (ed) 2013 *AC Electric Motors Control: Advanced Design Techniques and Applications* (New York: Wiley)

[8] Xia C L 2012 *Permanent Magnet Brushless DC Motor Drives and Control* (New York: Wiley)

[9] Krishnan R 2010 *Permanent Magnet Synchronous and Brushless DC Motor Drives* (Boca Raton, FL: CRC Press)

[10] Gieras J F, Wang R-J and Kamper M J 2008 *Axial Flux Permanent Magnet Brushless Machines* 2nd edn (Berlin: Springer)

Part III

Catalogue selection of motors

The correct selection of an electric motor that drives a mechanism is a matter of economic interest because it affects the capital and the running cost of the drive system, bearing in mind the enormous number of machines that employ electric motors throughout the industry. This part addresses major considerations for selecting an electric motor for specific working conditions, and the adaptation of engineering design solutions to standard industrial motor catalogues.

Chapter 7

Major considerations in selecting electric motors

The selection of an electric motor for a specified application requires, among other concerns, consideration of overloading, service life, ambient and cooling conditions. This chapter addresses motor power losses, motor life expectancy, and the thermal stress capacity of the motor [1–5].

7.1 Essential selection points

The correct selection of an electric motor to drive a given mechanism is imperative for the following reasons:

(a) Inadequate power capacity of the electric motor(s) might not address the intended functions of the working machine and would reduce its productivity. Moreover, the motor might experience premature deterioration that would reduce its life expectancy.

(b) An overdesigned motor exhibits higher losses, low efficiency, low power factor, and increases the capital cost of the entire system.

(c) The power capacity of the motor(s) should be determined conforming to the working machine requirements while maintaining its allowed temperature rise and within its power rating.

(d) Proper selection of power rating should address motor operation at steady state and transient conditions. To achieve that objective, a load diagram should be constructed. Load diagram means, torque or current or power as function of time.

(e) In addition to its steady state and transient conditions, motor operation might include short periods of overloading and starting or braking modes. Throughout all its working conditions, the motor should operate without overheating.

The thermal parameters are the fundamental criteria for selecting a motor because overheating is the major source of curtailing the service life of motors. Following the thermal evaluation, the engineer should also assess the capacity of the motor to endure overload events, current spikes, and peak torque demands.

7.2 Power losses in motors

Power loss with time (watt-second or joule) is the cause for temperature build up in electric motors. The sources for that power loss are hysteresis and turbulence currents in the iron core, ohmic losses in the copper conductors, windage and friction losses.

At the operating region of a motor, which is from the no-load speed to its rated speed, the iron, windage and friction losses can practically be considered constant. The total power losses would be:

$$\Delta P = K + P_{cu} \ [\text{W}] \tag{7.1}$$

where
    ΔP [W] is the total power loss in the motor,
    K [W] represents the constant power loss (iron, windage and friction), and
    P_{cu} [W] represents the variable power loss, which is in effect the copper losses.

At normal operating conditions, the input power to the motor can practically be expressed as follows:

$$P_{in} \cong K_1 \cdot I \ [\text{W}] \tag{7.2}$$

where
    P_{in} [W] is the input power,
    I [A] is the input current, and
    K_1 [V] is a constant value (for AC motors one can assume $PF \cong$ const.).

In practice, the total copper losses in the motor can be approximated as:

$$P_{cu} \cong K_2 \cdot I^2 \ [\text{W}] \tag{7.3}$$

where K_2 [Ω] is the equivalent resistance of the winding.
    From the above equations (equations (7.2) and (7.3)):

$$P_{cu} \cong K_2 \left(\frac{P_{in}}{K_1}\right)^2 = K_3 \cdot P_{in}^2 \ [\text{W}] \tag{7.4}$$

Following,

$$\left.\begin{array}{c} \dfrac{P_{cu}}{P_{cu,n}} = \dfrac{K_3 \cdot P_{in}^2}{K_3 \cdot P_{in,n}^2} \cong \left(\dfrac{P}{P_n}\right)^2 = x^2 \\[4mm] x = \dfrac{P}{P_n} \end{array}\right\} \tag{7.5}$$

where

$P_{cu,n}$ [W] is the variable power loss at rated load (at nominal shaft power),

$P_{in,n}$ [W] is the input power to the motor at rated load,

P [W] is the actual shaft power,

P_n [W] is the rated load (the nominal shaft power) of the motor, and

x is the normalized shaft power.

Finally, the total power loss (equation (7.1)) can now be expressed as follows:

$$\left.\begin{aligned} \Delta P = K + P_{cu} \cong K + P_{cu,n} \cdot x^2 = P_{cu,n}(a + x^2) \text{ [W]} \\ a = \frac{K}{P_{cu,n}} \end{aligned}\right\} \qquad (7.6)$$

where a is defined as the motor-loss coefficient.

Depending on the speed and design of the motor, the range of the motor-loss coefficient is [5]:

$$0.4 \lesssim a \lesssim 1.1$$

Note: In practice, the mechanical and friction losses are relatively very small compared with the motor iron losses and can be neglected.

7.3 Life expectancy of electric motors

Under normal operating conditions, that is within the manufacturer catalogue specifications, the life expectancy of an electric motor depends mainly on the thermal stress conditions. The state of the internal insulation determines the lifetime of the motor. Its design specifications include the life expectancy, which depends on the type of winding insulation. An increase in the operating temperature causes the quality of the insulation materials to deteriorate. The increase in heat would cause irreversible damage to the motor windings and might lead to leakage currents and to potential internal short circuits between turns.

An empirical equation developed by Arrhenius provides the life expectancy of the insulation at continuous elevated temperature. For example, stochastic analysis of laboratory tests on class H insulated motors offers the following Arrhenius equation [6]:

$$\gamma = 4.4825 \cdot 10^{-15} \cdot \exp\left(\frac{\dfrac{1.38}{0.8617 \cdot 10^{-4}}}{273.15° + \theta°}\right) \text{ [year]} \qquad (7.7)$$

where

γ [year] is the life expectancy of the motor, and

θ [°C] is the steady state temperature.

The behavior of the life expectancy as a function of the temperature (equation (7.7)) is presented by a semi-logarithmic graph in figure 7.1.

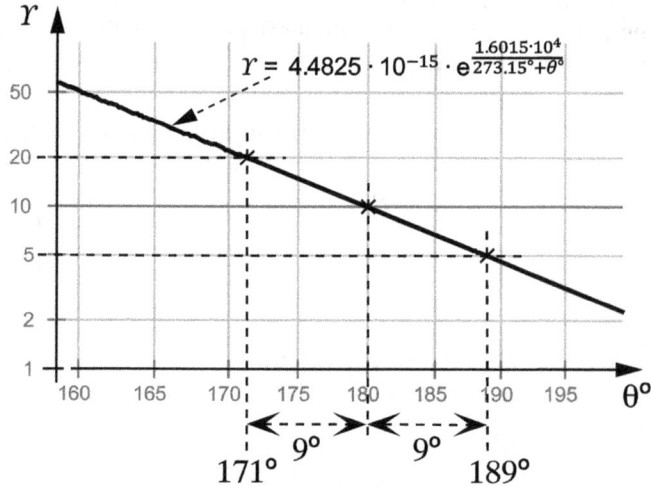

Figure 7.1. Service life versus temperature curve of a class H insulated motor.

A manufacturer design recommendation of a 10-year service life of a motor operating continuously at 180 °C would be increased or decreased by a factor of 2 when its temperature varies by ∓ 9 °C (figure 7.1). That is, a continuous operation at 171 °C would extend the service life to 20 years. But at 189 °C, the motor life would be reduced to only 5 years.

Different insulating materials may provide different slopes of the graph (figure 7.1), but the concept remains the same. That is, instead of ∓ 9 °C, it might be ∓ 8 °C or ∓ 10 °C that will extend or reduce the life expectancy of the motor by a factor of 2.

7.4 Motor heating and cooling processes

The total heat developed by the motor depends on its operating conditions and the direction of the internal heat flow. At no- or low-load, most of the heat is contributed by the iron core and transferred to the winding and to the remaining parts of the motor. Under loading conditions, the bulk amount of heat is generated by the winding conductors and then transfers to the iron core. Those heating processes are highly complicated to evaluate because of the intricate construction of the motor.

A simplified and yet practical assessment of the heat process uses the following assumptions:

    (a) The motor is a uniform body where all parts possess the same temperature. This assumption suggests infinite thermal conductivity of all motor parts.

(b) The rate of the heat transfer from motor to air, or to any other surrounding environment, depends linearly on the temperature difference between the two.

(c) Heat transfer by radiation is very limited and practically can be neglected.

Heat was measured by calories where 1 calorie is the quantity of heat required to raise the temperature of 1 g of water by 1 °C [7, 8]. The calorie is now replaced by the energy required to generate the heat (table 7.1).

Table 7.1. Conversion of heat quantities into energy units.

Heat quantity	Energy units
0.239 [calorie]	$\cong 1$ [joule $\equiv$ watt $\cdot$ s $\equiv$ newton·meter] $\cong 0.7376$ [ft $\cdot$ lbs]
252 [calorie]	$\cong 1$ [BTU—British Thermal Unit] $\cong 1054.8$ [joule]
860 [kilo-calorie]	$\cong 1$ [kWh—kilowatt hour] $= 3.6 \cdot 10^6$ [joule]

At a given load, the total heat generated by the motor is partly absorbed by the body of the motor itself and partly transfers to the surrounding environment. That is:

$$\left. \begin{array}{c} \text{Total generated heat } = \text{ Heat transfered to ambient } + \text{ Heat absorbed by motor} \\ \text{or} \\ \Delta P \cdot dt = A \cdot \theta \cdot dt + C \cdot d\theta \end{array} \right\} \quad (7.8)$$

where
  ΔP [W] is the total power loss in the motor,
  dt [s] is unit time,
  A [J s^{-1} °C^{-1}] is the heat transfer coefficient, which is the amount of heat transferred to the surrounding environment per unit time and per degree difference from ambient temperature,
  θ [°C] is the degree difference from ambient temperature,
  $\Delta\theta$ [°C] is the rate of change in temperature, and
  C [J °C^{-1}] is the heat capacity, which is the amount of heat required to raise the temperature of the motor by 1 °C.

Heat balance is achieved when the motor temperature reaches a steady (stable) maximum value, which indicates that the total amount of heat generated by the motor per unit time is transferred to the surrounding environment. It implies that the motor does not absorb additional heat. In equation (7.8), the rate of change in temperature would be $d\theta = 0$. As a result:

$$\Delta P \cdot dt = A \cdot \theta_{mx} \cdot dt \implies \theta_{mx} = \frac{\Delta P}{A} \left[\frac{\text{watt}}{\text{joule s}^{-1} \cdot {}^\circ \text{C}^{-1}} = {}^\circ \text{C} \right] \quad (7.9)$$

where θ_{mx} [°C] is a steady (stable) maximum temperature.

Heating process—The rise in motor temperature as a function of time can be deduced from equation (7.8):

$$\frac{d\theta}{dt} + \frac{A}{C} \cdot \theta = \frac{\Delta P}{C} \tag{7.10}$$

The solution of the first-order differential equation (equation (7.10)) is:

Heating
$$\left. \begin{array}{c} \theta(t) = \theta_{mx} \left(1 - e^{-\frac{t}{\tau}}\right) + \theta_0 \cdot e^{-\frac{t}{\tau}} \ [°C] \\[4pt] \text{where: a final (high) temperature is: } \theta_{mx} \ [°C] \\[2pt] \text{an initial (low) temperature is: } \theta_0 \ [°C] \\[4pt] \text{the thermal time constant is: } \tau = \frac{C}{A} \ [s] \end{array} \right\} \tag{7.11}$$

Cooling process—The decrease in motor temperature as a function of time can be deduced from equation (7.11). In this case, the initial temperature is at a peak value θ_{mx}, and it decays in time to a lower value θ_0:

Cooling
$$\left. \begin{array}{c} \theta(t) = \theta_0 \left(1 - e^{-\frac{t}{\tau}}\right) + \theta_{mx} \cdot e^{-\frac{t}{\tau}} \ [°C] \\[4pt] \text{where: a final (low) temperature is: } \theta_0 \ [°C] \\[2pt] \text{an initial (high) temperature is: } \theta_{mx} \ [°C] \\[4pt] \text{the thermal time constant is: } \tau = \frac{C}{A} \ [s] \end{array} \right\} \tag{7.12}$$

The temperature as a function of time during heating and cooling (equatons (7.11) and (7.12)) is presented graphically in figure 7.2.

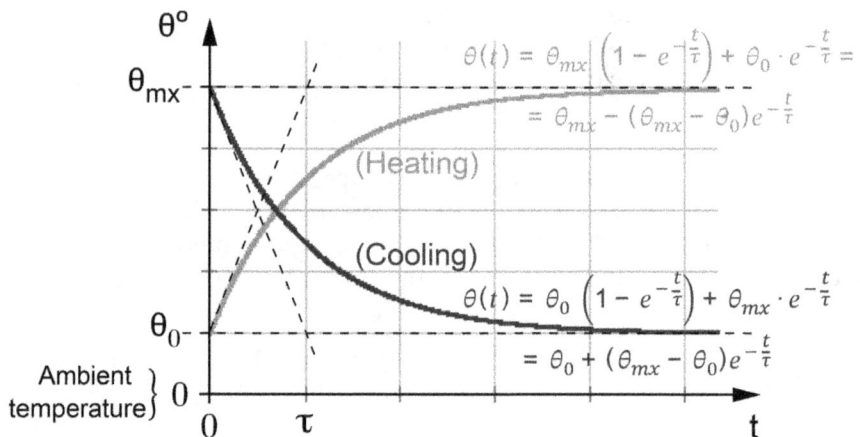

Figure 7.2. Heating and cooling curves as function of time.

7-6

Experimental tests show that the simplified exponential curve(s) derived above (figure 7.2) is a near match with the actual thermal behavior of a motor [5].

Notes:

(1) Zero temperature in figure 7.2 denotes the ambient temperature (not the temperature at which water freezes).

(2) The thermal time constant τ [s] depends on the cooling conditions of the motor:

   With *external cooling or forced cooling*, such as in methods of forced air or oil cooling techniques, the thermal time constant remains the same at all operating conditions of the motor; i.e., the heat transfer coefficient is not affected by the working condition of the motor, that is: $A = $ const.

   With *self-cooling or self-ventilating*, such as when an internal fan is mounted on the motor shaft, the thermal time constant varies between starting (acceleration), braking (deceleration), and rest (standstill) conditions, that is: $A \neq$ const. (addressed in section 8.5.2).

(3) To prevent excess decay in the motor insulation, motor temperature should not exceed catalogue ratings. Motor manufacturers usually provide a service factor that allows continuous operation of the motor beyond its rated power for a short period of time, that is within the motor thermal time constant.

Power loss depends on the working conditions of the motor. As the load varies, the maximum temperature of the motor varies too (figure 7.3).

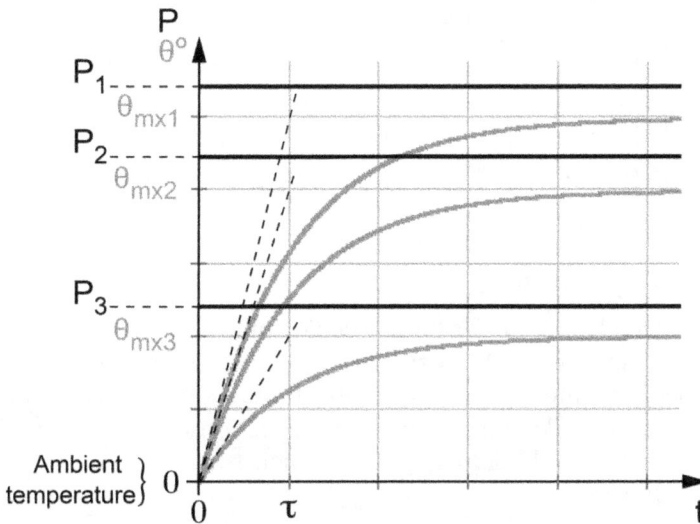

Figure 7.3. Heating curves at different motor loads.

Continuous operation of a motor at different load conditions, $P_1 > P_2 > P_3$, brings about different maximum temperatures, $\theta_{mx1} > \theta_{mx2} > \theta_{mx3}$ (figure 7.3).

Variations in load power would bring about variations in motor losses, and the maximum temperature would change too. In contrast, the thermal time constant τ [s] remains unchanged because it is not a function of the motor load (equation (7.11)).

Example 7.1 An induction motor operates at rated load. The motor has the following parameters: rated power $P_n = 100$ kW, rated efficiency $\eta_n = 0.92$, overall weight $G = 800$ kg, a combined thermal capacity $C_G = 0.48$ kJ (kg · °C) $-$ 1, and a maximum temperature rise $\theta_{mx} = 50$ °C.

Calculate the thermal time constant of the motor.

Solution

The motor efficiency is given by:

$$\eta_n = \frac{P_n}{P_{inn}} = \frac{P_n}{P_n + \Delta P_n}$$

where

$P_{in,n}$ is the input power to the motor at rated load, and
ΔP_n is the total power loss at rated load.

Using the above equation, the total power loss is:

$$\Delta P_n = P_n\left(\frac{1}{\eta_n} - 1\right) = 100\left(\frac{1}{0.92} - 1\right) = 8.7 \ [\text{kW}]$$

And the thermal time constant τ is:

$$\left.\begin{array}{c} \tau = \dfrac{C}{A} \\ \text{divided by} \\ \theta_{mx} = \dfrac{\Delta P_n}{A} \end{array}\right\} \tau = \frac{C \cdot \theta_{mx}}{\Delta P_n} = \frac{(0.48 \cdot 800) \cdot 50}{8.7} = 2{,}207 \text{ s} = 36.8 \ [\text{min}].$$

7.5 Modes of operation

The selection of an electric motor for a specific application depends on the working conditions of the mechanism. Four types of working conditions are addressed here: continuous operation mode, short-time operation mode, intermittent cycle operation mode, and random operation. In the four sections below, it was assumed that the motor has an external cooling system, which suggests the same thermal time constant τ for all cases.

7.5.1 Continuous operation mode

In a continuous working condition or at slightly variable load, the motor operates at a given load for a long period of time, much longer than its thermal time constant. That way, the motor temperature reaches a stable value (heat balance).

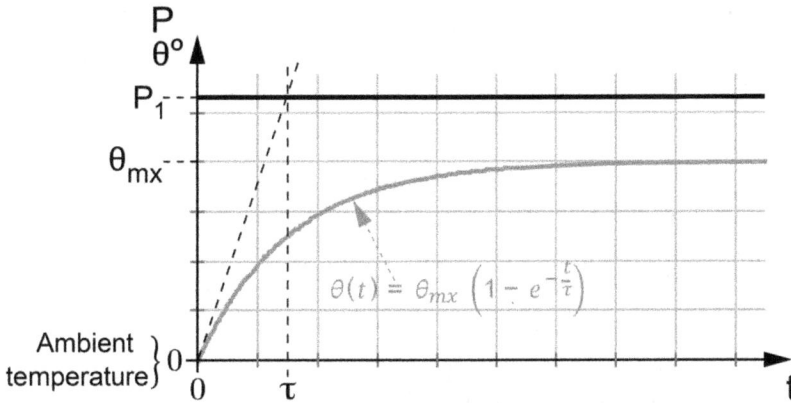

Figure 7.4. Heating curve at continuous mode of operation.

As shown (figure 7.4), the motor operates at a continuous power P_1. Motor heating begins at ambient temperature, $\theta_0 = 0$. In practice, after three or four thermal time constants, τ, heat balance is achieved and the working motor reaches a stable temperature θ_{mx}. Using equation (7.11), the temperature behavior as a function of time is:

$$\theta(t) = \theta_{mx}\left(1 - e^{-\frac{t}{\tau}}\right) \tag{7.13}$$

Examples of continuous mode of operation are pumps and fans.

7.5.2 Short-time operation mode

A short-time working condition means that the operating time (duty time) is in the order of the thermal time constant of the motor. Consequently, heat balance would not be achieved, and the temperature would not reach a stable value during motor operation. Following that duty time, the motor is allowed to cool down to ambient temperature.

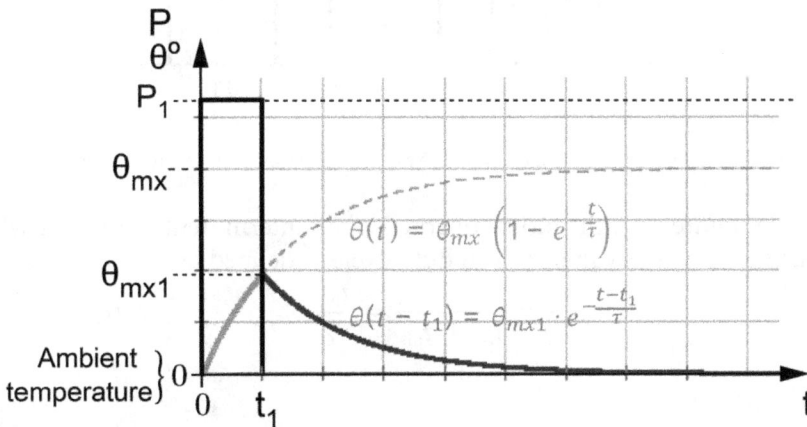

Figure 7.5. Heating and cooling curves during short-time operation.

As shown (figure 7.5), the motor operates at a power P_1 for a short period of time t_1. Motor heating begins at ambient temperature, $\theta_0 = 0$, and increases up to a peak value θ_{mx1} at t_1. Using equation (7.11), the temperature behavior as a function of time is:

$$\theta(t) = \theta_{mx}\left(1 - e^{-\frac{t}{\tau}}\right) \ [°C] \tag{7.14}$$

where θ_{mx} [°C] is the maximum temperature that the motor would have reached as if operating continuously at P_1 [W].

At $t = t_1$, the motor is disconnected and its temperature drops to ambient value. Using equation (7.12), the temperature behavior as a function of time is:

$$\theta(t - t_1) = \theta_{mx1} \cdot e^{-\frac{t-t_1}{\tau}} \ [°C] \tag{7.15}$$

where θ_{mx1} [°C] is the maximum temperature at $t = t_1$.

Examples of short-time operating mode are a draw bridge actuator and a starter of a car engine.

7.5.3 Intermittent cycle operation mode

A repetitive operation of the motor at a given duty cycle is called an intermittent mode of operation (figure 7.6). At each operating cycle, the duty time t_{on} is in the order of the motor thermal time constant. Heat balance is not achieved, and the temperature does not reach a stable value. During the relatively short downtime t_{off}, the motor temperature does not cool down to ambient value.

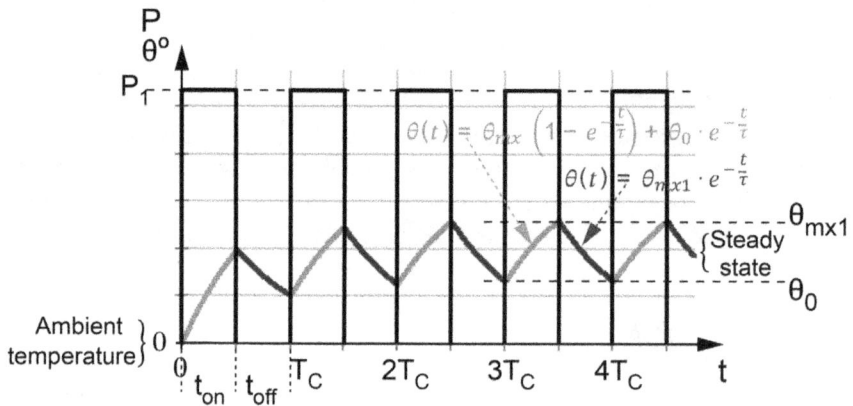

Figure 7.6. Heating and cooling curves during intermittent operation mode.

As shown (figure 7.6), the motor operates at an intermittent duty cycle where the power during each interval is P_1. The duty cycle is defined as:

$$\varepsilon = \frac{t_{on}}{t_{on} + t_{off}} = \frac{t_{on}}{T_C} \tag{7.16}$$

where
  ε is the duty cycle,
  t_{on} is the duty time at each operating cycle,
  t_{off} is the downtime at each operating cycle, and
  $T_C = t_{on} + t_{off}$ is the cycle time.

Following a few cycles of operation, the fluctuation in temperature reaches a steady state condition where the minimum temperature is θ_0 and the peak temperature is θ_{mx1}.

Consider one cycle during the steady state condition, the temperature rise as a function of time is (equation (7.11)):

$$\theta(t) = \theta_{mx} \left(1 - e^{-\frac{t}{\tau}}\right) + \theta_0 \cdot e^{-\frac{t}{\tau}} \ [°C] \tag{7.17}$$

where

θ_{mx} [°C] is the maximum temperature that the motor would have reached as if operating continuously at P_1 [W], and

θ_0 [°C] is the initial temperature at each steady state thermal cycle.

The cooling behavior as a function of time is (equation (7.12)):

$$\theta(t) = \theta_{mx1} \cdot e^{-\frac{t}{\tau}} \ [°C] \tag{7.18}$$

where θ_{mx1} [°C] is the maximum temperature at each steady state thermal cycle.

Examples of intermittent mode of operation are refrigeration systems and elevators.

7.5.4 Random operation

As the title suggests, the operating condition of the motor is totally arbitrary. That is, a random operation without method or pattern (figure 7.7).

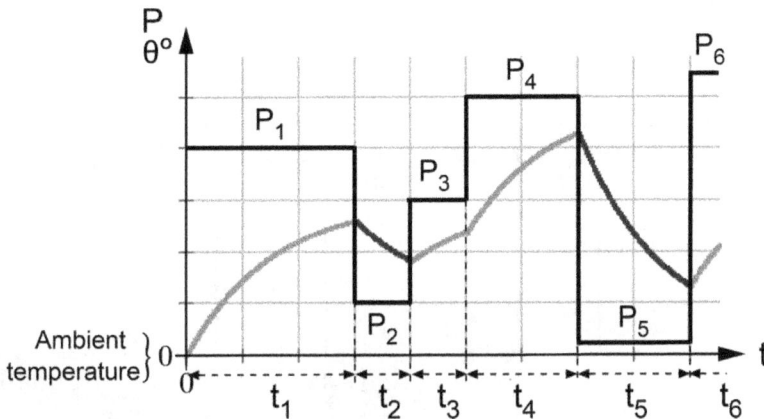

Figure 7.7. Heating and cooling curves during random operation.

As shown (figure 7.7), as the motor power increases, its temperature increases. As the motor power decreases, its temperature cools down.

7.6 Problems

1. An electric motor operates continuously at its rated load P_n. The motor heating characteristic $\theta = f(t)$ at that load is given in figure P7.1.

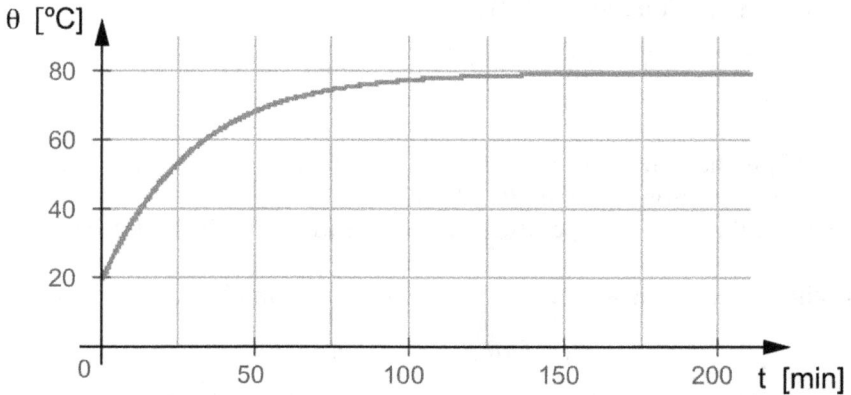

Figure P7.1. Motor heating characteristic at P_n continuous motor load.

The motor load is reduced to $0.45 \cdot P_n$ (continuous operation). Assume that the motor constant losses equal its variable losses at rated power, and calculate the maximum temperature at that reduced load.

2. A self-ventilated motor operates at an intermittent cycle mode. The cycle time is $T_C = 70$ min. The duty cycle is $\varepsilon = 4/7$ (figure P7.2). The thermal time constant during the duty cycle is 25 min, and during off-time is 50 min.

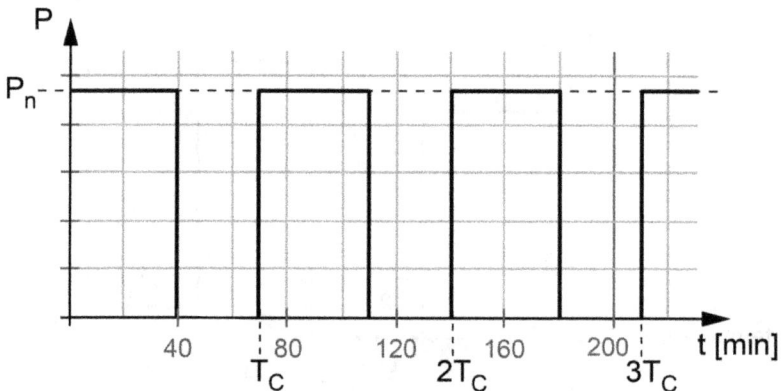

Figure P7.2. An intermittent cycle mode where the motor turns on and off.

When the motor operates continuously at rated load P_n, its temperature reaches a maximum of 90 °C above ambient.

Calculate the lower and upper temperatures at steady state operation.

3. The same electromechanical drive as in problem 2. But during off-time, the mechanical load is detached from the motor shaft, and the motor continues to run at no load (figure P7.3). The cycle time is $T_C = 70$ min. The duty cycle is $\varepsilon = 4/7$.

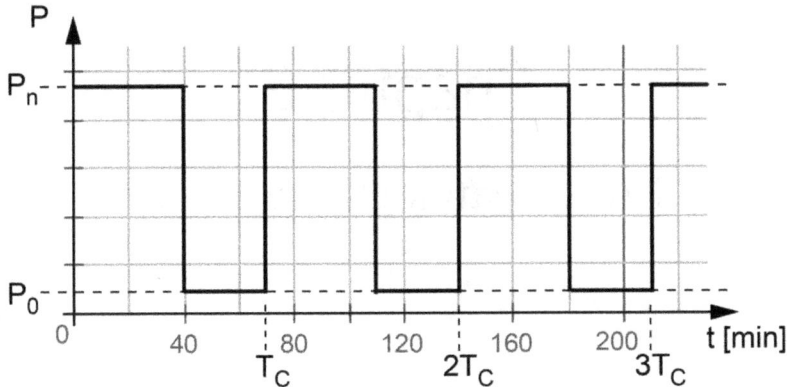

Figure P7.3. An intermittent cycle mode where the motor rotates at off-time.

The thermal time constant during the duty time is 25 min. During off-time, the motor rotates a bit faster (better ventilation) and the thermal time constant is 20 min.

When the motor operates continuously at no load P_0, its temperature reaches a maximum of 10 °C above ambient. When it operates continuously at rated load P_n, its temperature reaches a maximum of 90 °C above ambient.

Calculate the lower and upper temperatures at steady state operation.

References

[1] Mohan N 2001 *Electric Drives: An Integrative Approach* (MNPERE Publisher)
[2] Dubey G K 2001 *Fundamentals of Electrical Drives* (Alpha Science International Ltd)
[3] Nasar S A and Unnewehr L E 1983 *Electromechanical and Electric Machines* (New York: Wiley)
[4] Meyers R A (ed) 2002 *Encyclopedia of Physical Science and Technology* **vol 5** 3rd edn (New York: Academic)
[5] Chilikin M 1976 *Electric Drive* (Moscow: MIR Publishers)
[6] Brancato E L 1992 Estimation of lifetime expectancies of motors *IEEE Electr. Insul. Mag.* **8** 5–13
[7] Ronnie R and Law J (ed) 2019 *A Dictionary of Physics* 8th edn (Oxford: Oxford University Press)
[8] Fink D G and Beaty H W (ed) 2000 *Standard Handbook for Electrical Engineers* 14th edn (New York: McGraw-Hill)

Chapter 8

Motor catalogue selection

Standard motor catalogues are available for a variety of duty cycles such as for continuous, for short-time, and for intermittent periodic operations. Among many considerations, an inadequate power capacity might cause premature deterioration of materials and shorten the life expectancy of the motor. This chapter addresses design procedures that include the steady-state as well as the transient conditions of the electromechanical drive system, the construction of load diagrams, the effect of ambient temperatures, and finally, the selection of the appropriate motor from standard manufacturer catalogues [1–4].

8.1 Catalogue selection at continuous and random operating modes

Continuous operation at constant or at slightly variable load, brings the motor temperature to a maximum steady-state level (see section 7.5.1). In this case, a quantified average shaft power can be used for a catalogue search. As such, the selection becomes a relatively simple task. That is:

$$\left.\begin{array}{c} P_{\mathrm{m}} = T_m \cdot \omega_m \ [\mathrm{W}] \\ \text{and from a continuous operation catalogue, select:} \\ P_n^{(cat)} \geqslant P_m \ [\mathrm{W}] \end{array}\right\} \qquad (8.1)$$

where

P_{m} [W] is the quantified (average) shaft power,
T_m [Nm] is the load torque referred to the motor shaft (section 2.1),
ω_m [s^{-1}] the radial speed of the motor shaft, and
$P_n^{(cat)}$ [W] is the catalogue (nameplate) power.

Note: In a continuous operating mode, power losses during startup can be neglected because those losses practically have no effect on the motor heating.

Once the required motor power P_m is quantified, the catalogue selection P_{cat} should have equal or slightly higher power rating (equation (8.1)). That would assure maximal use of material, better motor efficiency, and would lower its capital cost.

In addition, one must adapt the selected motor to the actual ambient temperature (see section 8.2), and also check whether the motor is capable of addressing its startup and the maximum torque demand. If not adequate, a motor with higher rated power should be considered.

Random operation means that the mechanism operates without a method or a pattern (section 7.5.4). Therefore, a conservative selection must be applied. That is, one should choose the higher load demand as if the motor operates continuously at that power level.

8.2 Ambient temperature consideration

Motor manufacturers design electric motors with reference to a standard ambient temperature specified in their catalogue; for example: a standard temperature of 40 °C is suggested in ANSI/NEMA Standards [4]. In practice, motors are used on many locations where the ambient temperature differs from the standard, such as in a cold refrigeration room or in a hot machine room. To account for a variation from the standard ambient temperature, a re-rating or de-rating of the motor nameplate power is required.

The rated temperature rise of a motor operating continuously at its nominal (nameplate) power is:

$$\theta_{mx,n} = \theta_{mx}^{(cat)} - \theta_0^{(cat)} \ [°C] \tag{8.2}$$

where

$\theta_{mx,n}$ [°C] is the rated rise in temperature,

$\theta_{mx}^{(cat)}$ [°C] is the nominal (catalogue) maximum temperature, and

$\theta_0^{(cat)}$ [°C] is the nominal (catalogue) ambient temperature.

The rated temperature rise (equation (8.2)) can also be expressed by the total power losses when the motor operates at its nominal power (equations (7.6) and (7.9)):

$$\theta_{mx,n} = \theta_{mx}^{(cat)} - \theta_0^{(cat)} = \frac{\Delta P_n}{A} = \frac{P_{cu,n}(a + 1^2)}{A} \ [°C] \tag{8.3}$$

where

ΔP_n [W] is the total motor power loss at rated load P_n. In this case, the normalized shaft power (equation (7.5)) is: $x = 1$, and

A [J s^{-1} °C^{-1}] is the heat transfer coefficient (see equation (7.8)).

When the ambient temperature differs from the standard, the rated maximum rise in temperature must be modified:

$$\theta_{mx,u} = \theta_{mx,n} - \Delta\theta \ [°C] \tag{8.4}$$

where

$\theta_{mx,u}$ [°C] is the modified (updated) rise in temperature, and

$\Delta\theta$ [°C] is the difference between the actual ambient temperature and the catalogue (standard) value.

Applying the power loss formulas (equations (7.6) and (7.9)) when the motor operates at a shaft power different than its nominal value, one gets:

$$\theta_{mx,u} = \theta_{mx,n} - \Delta\theta = \frac{\Delta P_u}{A} = \frac{P_{cu,n}(a + x^2)}{A} \; [°C] \tag{8.5}$$

where ΔP_u [W] is the power loss when the motor operates at a shaft power different than its nominal value.

The ratio of the updated rise in temperature (equation (8.5)) over the rated temperature rise (equation (8.3)) suggests:

$$\left. \begin{array}{c} \dfrac{\theta_{mx,u}}{\theta_{mx,n}} = \dfrac{\theta_{mx,n} - \Delta\theta}{\theta_{mx,n}} = \dfrac{a + x^2}{a + 1} \\[2mm] \text{and the value of } x \text{ becomes:} \\[2mm] x = \sqrt{1 - \dfrac{\Delta\theta}{\theta_{mx,n}}(a + 1)} \end{array} \right\} \tag{8.6}$$

where

x is the normalized shaft power (equation (7.5)), and

a is the motor loss coefficient (equation (7.6)).

Finally, the re-rated or de-rated nameplate power of the motor is:

$$P_x = x \cdot P_n^{(cat)} \; [W] \tag{8.7}$$

where

P_x [W] is the re-rated/de-rated power of the motor, and

$P_n^{(cat)}$ [W] is the motor nameplate (catalogue) power.

Example 8.1 The nameplate power of an electric motor is 15 kW. Its nominal (catalogue) ambient temperature is $\theta_0^{(cat)} = 40\,°C$, and its maximum allowed (catalogue) temperature is $\theta_{mx}^{(cat)} = 105\,°C$.

Assume that the constant power loss of the motor equals its variable power loss at rated load, and:

(a) Re-rate the nominal power of the motor at an ambient temperature of 30 °C.

(b) De-rate the nominal power of the motor at an ambient temperature of 50 °C.

Solution

(a) The difference in ambient temperature is: $\Delta\theta = 30 - 40 = -10\,°C$

The rated rise in temperature is (equation (8.2)):

$$\theta_{mx,n} = \theta_{mx}^{(cat)} - \theta_0^{(cat)} = 105 - 40 = 65\,°C$$

The motor-loss coefficient is (equation (7.6)): $a = 1$

The normalized shaft power is (equation (8.6)):

$$x = \sqrt{1 - \frac{-10}{65}(1 + 1)} = 1.144$$

And the re-rated power of the motor at the lower ambient temperature is (equation (8.7)):

$$P_x = 1.144 \cdot 15 = 17.2\ [\text{kW}]$$

(b) The normalized shaft power is:

$$x = \sqrt{1 - \frac{50 - 40}{65}(1 + 1)} = 0.832$$

And the de-rated power of the motor at the higher ambient temperature is:

$$P_x = 0.832 \cdot 15 = 12.5\ [\text{kW}]$$

8.3 Motor selection at periodic variable-load-operating mode

Motor selection at continual operating mode where the load varies substantially must factor in fluctuation in temperature. As the load varies, so does the motor temperature. An increase in load brings about higher power losses that cause the motor temperature to rise. A reduction in the load brings down the losses which leads to a decrease in the motor temperature.

8.3.1 Load diagram

The first step in selecting a motor for driving a mechanism is to construct a load diagram. That is to devise a diagram of the required shaft power or torque, or of the motor input current as a function of time.

An arbitrary load diagram (figure 8.1) shows a periodic fluctuation of motor shaft power as a function of time. At each time-interval t_i, the mechanism load power referred to the motor shaft, P_i [W], has to be calculated (see chapter 2). At the first time-interval t_1, the referred load power is P_1. At t_2, the calculated shaft power is P_2, and so on. At the end of the nth time-interval t_n, the power-cycle repeat itself. The cycle time is:

$$T_C = \sum_{i=1}^{n} t_i \tag{8.8}$$

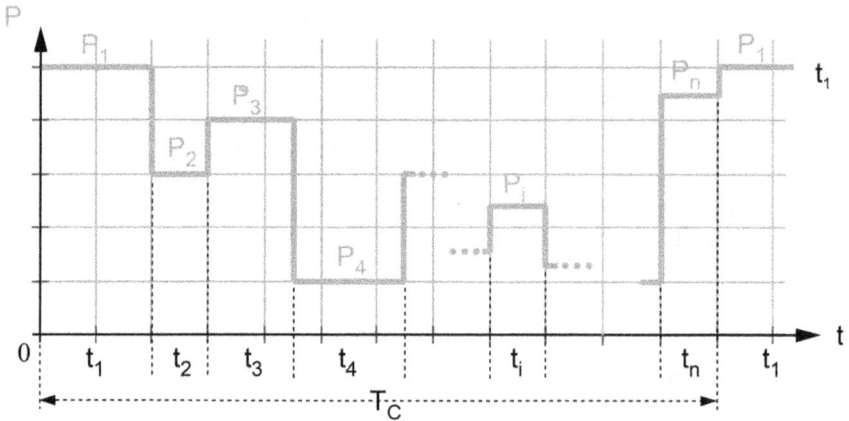

Figure 8.1. Load diagram of motor shaft power.

where
> T_C is the cycle time,
> i is the summation index denoting the time-interval number,
> n is the total number of time-intervals in one cycle, and
> t_i is the interval time.

Note: For an operation at a non-cyclic fluctuated load, see note #1 that addresses equation (8.21).

8.3.2 Average power loss method

Using a calculated average power in the load diagram of figure 8.1 as a criterion for motor selection would be the wrong engineering approach because the power losses, not the power demand, is the cause for the temperature rise of the motor. The right engineering procedure would be to construct a power loss diagram that is based on the power demand diagram. Following, a calculated average power loss would be used as the criterion for the motor selection.

   The method of average power loss of a periodic variable load requires first to calculate the average loss at each time-interval of the load diagram (figure 8.1). Then, an average power loss can be calculated and used to select the power of the motor itself. That process would assure that the motor would operate within its designed temperature boundaries. Figure 8.2(a) shows the load diagram at the motor shaft (same as in figure 8.1), the calculated power loss at each time-interval (dashed line). Figure 8.2(b) shows the rise and fall of the motor temperature during its cyclic operation.

   The power loss at each time-interval (figure 8.2(a)) can be derived from the definition of the motor efficiency:

$$\left. \begin{aligned} \eta_i &= \frac{P_i}{P_i + \Delta P_i} \\ \text{and: } \Delta P_i &= P_i \left(\frac{1}{\eta_i} - 1 \right) \end{aligned} \right\}$$

(8.9)

where

P$_i$ [W] is the shaft power at interval t_i,
ΔP_i [W] is the power loss of the motor at interval t_i, and
η_i is the motor efficiency at P_i.

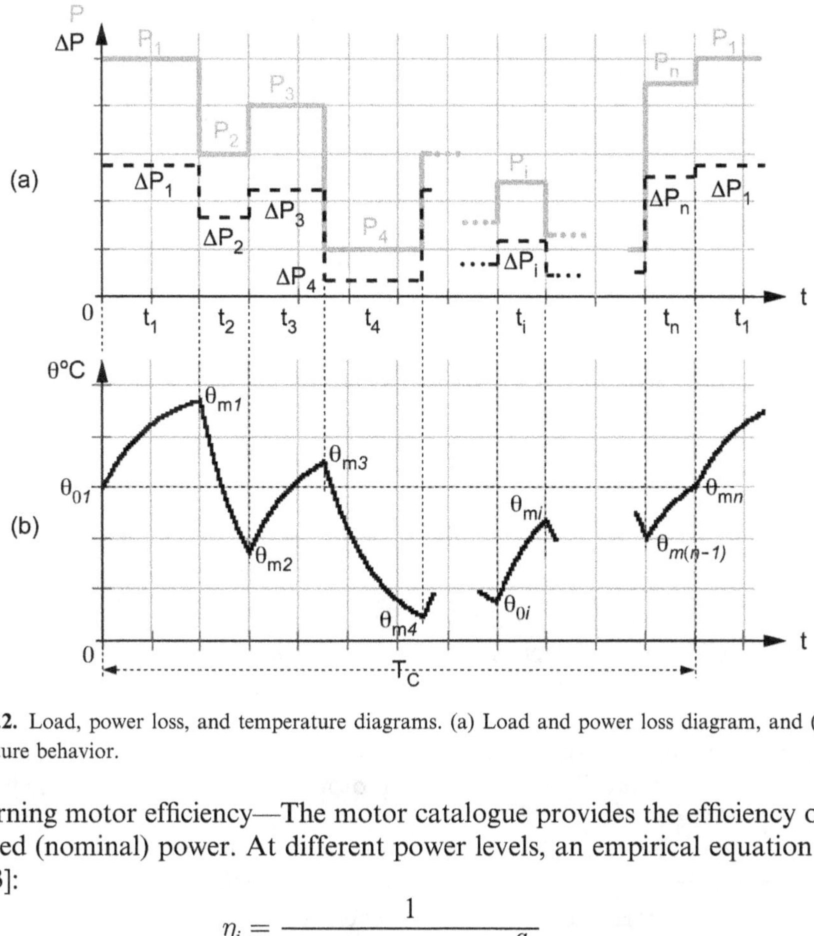

Figure 8.2. Load, power loss, and temperature diagrams. (a) Load and power loss diagram, and (b) motor temperature behavior.

Concerning motor efficiency—The motor catalogue provides the efficiency only for the rated (nominal) power. At different power levels, an empirical equation can be used [3]:

$$\eta_i = \cfrac{1}{1 + \left(\cfrac{1}{\eta_n} - 1\right)\cfrac{x + \cfrac{a}{x}}{1 + a}} \tag{8.10}$$

where

η_n is the motor rated efficiency,
x is the normalized shaft power (equation (7.5)) at each time-interval i, and
a is the motor loss coefficient (equation (7.6)).

The behavior of the empirical motor efficiency (equation (8.10)) as a function of the normalized shaft power is as in figure 8.3.

As an example, assume that the constant power loss K equals to the variable power losses at rated power $P_{cu,\,n}$, which suggests that the motor loss coefficient is:

$a = 1$ (section 7.2). Also assume that the rated efficiency is $\eta_n = 0.92$. Then, the efficiency at any given shaft power can be ascertained from the graph (figure 8.3).

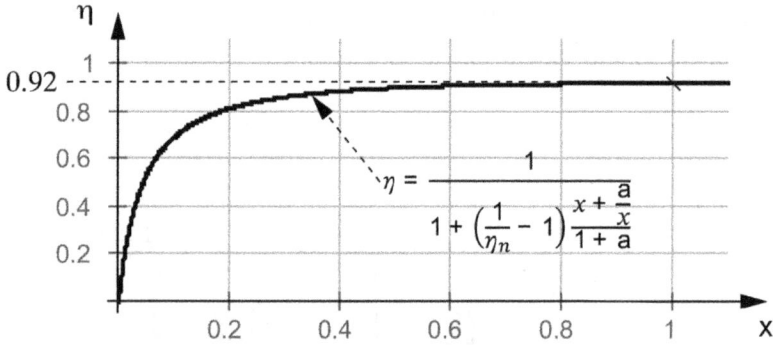

Figure 8.3. Motor efficiency as a function of the normalized shaft power where $a = 1$ and $\eta_n = 0.92$.

In equation (8.10), to obtain the motor loss coefficient and the rated efficiency, an estimated motor power should be pre-selected. This can be done by estimating a motor power directly from the load diagram (figure 8.2(a)):

$$P_{est} \cong \frac{1.2}{T_C} \sum_{i=1}^{n} P_i \cdot t_i \ [\text{W}] \tag{8.11}$$

where
P_{est} is an estimated power of the motor,
P_i is the shaft power at each time-interval (figure 8.1), and
1.2 is a coefficient of safety.

Following, an estimated motor of P_{est} with and efficiency η_{est} can be selected from a manufacturer catalogue. That estimated motor power and its efficiency is the starting step of the design.

In figure 8.2, the rise and fall in motor temperature $\theta°$ relate to the power loss ΔP_i at each time-interval t_i. The symbol m denotes a minimum and a maximum temperature at the start and at the end of each time-interval. Assuming the same thermal time constant throughout the cycle time T_C, the temperature as a function of time would be (see equations (7.11) and (7.12))

$$\theta_{mi} = \theta_{mx, i} \left(1 - e^{-\frac{t_i}{\tau}}\right) + \theta_{0i} \cdot e^{-\frac{t_i}{\tau}} \ [°\text{C}] \tag{8.12}$$

where
θ_{0i} is the temperature at the beginning of a time-interval and
$\theta_{mx, i} = \Delta P_i / A$ is the maximum temperature that the motor would have reached if it was operating at P_i continuously.

The temperature at the end of each time-interval can be expressed in a successive order:

$$\left.\begin{aligned}
\theta_{m1} &= \frac{\Delta P_1}{A}\left(1 - e^{-\frac{t_1}{\tau}}\right) + \theta_{01} \cdot e^{-\frac{t_1}{\tau}} \\
\theta_{m2} &= \frac{\Delta P_2}{A}\left(1 - e^{-\frac{t_2}{\tau}}\right) + \theta_{m1} \cdot e^{-\frac{t_2}{\tau}} \\
\theta_{m3} &= \frac{\Delta P_3}{A}\left(1 - e^{-\frac{t_3}{\tau}}\right) + \theta_{m2} \cdot e^{-\frac{t_3}{\tau}} \\
&\quad\vdots \quad \vdots \quad \vdots \\
\theta_{mn} &= \frac{\Delta P_n}{A}\left(1 - e^{-\frac{t_n}{\tau}}\right) + \theta_{m2} \cdot e^{-\frac{t_n}{\tau}}
\end{aligned}\right\} \tag{8.13}$$

Substituting θ_{m1} into the equation of θ_{m2}; then, substituting θ_{m2} into the equation of θ_{m3}, and so forth and so on, one can obtain the temperature at the end of the cycle:

$$\left.\begin{aligned}
\theta_{mn} &= \frac{\Delta P_n}{A}\left(1 - e^{-\frac{t_n}{\tau}}\right) + \frac{\Delta P_{n-1}}{A}\left(1 - e^{-\frac{t_{n-1}}{\tau}}\right)e^{-\frac{t_n}{\tau}} + \cdots \\
&+ \frac{\Delta P_2}{A}\left(1 - e^{-\frac{t_2}{\tau}}\right)e^{-\frac{T_C - (t_2 + t_1)}{\tau}} + \frac{\Delta P_1}{A}\left(1 - e^{-\frac{t_1}{\tau}}\right)e^{-\frac{T_C - t_1}{\tau}} \\
&+ \theta_{01}e^{-\frac{T_C}{\tau}}
\end{aligned}\right\} \tag{8.14}$$

The temperature at the beginning and the end of the cycle is the same:

$$\theta_{01} = \theta_{mn} \quad [°C] \tag{8.15}$$

Substituting θ_{01} (equation (8.15)) into equation (8.14), the temperature behavior becomes:

$$\left.\begin{aligned}
\theta_{mn}\left(1 - e^{-\frac{T_C}{\tau}}\right) &= \frac{\Delta P_n}{A}\left(1 - e^{-\frac{t_n}{\tau}}\right) + \frac{\Delta P_{n-1}}{A}\left(1 - e^{-\frac{t_{n-1}}{\tau}}\right)e^{-\frac{t_n}{\tau}} + \cdots \\
&+ \frac{\Delta P_2}{A}\left(1 - e^{-\frac{t_2}{\tau}}\right)e^{-\frac{T_C - (t_2 + t_1)}{\tau}} + \frac{\Delta P_1}{A}\left(1 - e^{-\frac{t_1}{\tau}}\right)e^{-\frac{T_C - t_1}{\tau}}
\end{aligned}\right\} \tag{8.16}$$

In practice, the cycle time T_C is much longer than the thermal time constant τ. Consequently, the left term (equation (8.16)) suggests that when the motor operates continuously at an average power loss ΔP_{avg}, its temperature would reach a steady-state peak value of θ_{mn}:

$$\theta_{mn} = \frac{\Delta P_{avg}}{A} \quad [°C] \tag{8.17}$$

The exponential terms in equation (8.16) can be presented as a sum of a series [5]:

$$\left.\begin{aligned}
e^{-x} &= 1 - x + \frac{x^2}{2!} - \frac{x^3}{3!} + \frac{x^4}{4!} + \cdots + \frac{x^n}{n!} \\
\text{For } x \ll 1, \text{ the exponential function} &\text{ can be approximated:} \\
e^{-x} &\cong 1 - x
\end{aligned}\right\} \tag{8.18}$$

Using the approximation above (equation (8.18)), one can derive the following expressions:

$$\left.\begin{array}{c} (1 - e^{-x_n}) \cong x_n \\ \text{and} \\ (1 - e^{-x_{n-1}})e^{-x_n} \cong x_{n-1} \end{array}\right\} \tag{8.19}$$

Substituting the definition of θ_{mn} (equation (8.17)) into the thermal equation (equation (8.16)), and applying the exponential approximations of equation (8.19), one gets:

$$\left.\begin{array}{c} \dfrac{\Delta P_{avg}}{A}\dfrac{T_C}{\tau} \cong \dfrac{\Delta P_n}{A}\dfrac{t_n}{\tau} + \dfrac{\Delta P_{n-1}}{A}\dfrac{t_{n-1}}{\tau} + \cdots + \dfrac{\Delta P_2}{A}\dfrac{t_2}{\tau} + \dfrac{\Delta P_1}{A}\dfrac{t_1}{\tau} \\ \text{and the average power loss is:} \\ \Delta P_{avg} \cong \dfrac{1}{T_C}\sum_{i=1}^{n}\Delta P_i \cdot t_i \ [\text{W}] \end{array}\right\} \tag{8.20}$$

Finally, the average motor power that would assure operation within the safe temperature boundaries is:

$$P_{avg} \geqq \dfrac{\Delta P_{avg}}{\dfrac{1}{\eta_{est}} - 1} \ [\text{W}] \tag{8.21}$$

where

P_{avg} is the approximated motor power,

ΔP_{avg} is the approximated power loss in the motor (equation (8.20)), and

η_{est} is the efficiency of the estimated motor power.

Notes:
(1) The average power formula (equation (8.21)) can be applied to a non-periodic operation provided that the motor is assigned to a specific load diagram. Otherwise, a random operating mode should be applied (section 8.1).
(2) An analysis of average power loss where the thermal time constant varies is addressed in section 8.5.2.

In summation, the design procedure of the average power loss method began with a selection of an estimated motor size P_{est} (equation (8.11)) to obtain an efficiency value η_{est} from a motor catalogue. Then, a power loss diagram $\Delta P = f(t)$ was constructed (figure 8.2(a)), and an average motor power was calculated (equation (8.21)). Still, if that calculated average power differs from the initial estimated motor size, the whole design procedure must be repeated by estimating a new motor power. Reiterating, the design procedure steps are:

Step 1: Construct a load diagram (figure 8.1).

Step 2: Estimate a motor power, P_{est} (equation (8.11)).

Step 3: Obtain the catalogue efficiency of the estimated motor power.

Step 4: Draw the efficiency curve (equation (8.10)).
Step 5: Construct a power loss diagram (equation (8.9)).
Step 6: Calculate the average power loss of the motor (equation (8.20)).
Step 7: Calculate the average power of the motor, P_{avg} (equation (8.21)).
Step 8: Compare the average power of Step 7 with the estimated motor of Step 2.
Step 9: If $P_{avg} \cong P_{est}$, the selection is complete. If not, go back to Step 2.

As suggested before, the selected motor based on the average loss ΔP_{avg} (equation (8.20)) must also be assessed for ambient temperature (section 8.2) and for starting and maximum torques. If not adequate, a motor with higher rated power should be considered.

Note: The above intricated average power loss method is a time-consuming job and can and should be computerized. Alternatively, a few hand calculation methods are proposed below.

8.3.3 Hand calculations—equivalent current method

The equivalent current method is based on the fact that a load diagram of the input current to the motor is available. Then, the method suggests to replace that load diagram with an equivalent current that would produce equal power losses in the operating motor.

Assuming:

(a) constant input voltage to the motor, and
(b) constant internal resistance.

Then, the total losses inside the motor would be (see equation (7.1) and (7.3)):

$$\Delta P = K + K_2 \cdot I^2 \ [\text{W}] \tag{8.22}$$

Substituting the above power loss (equation (8.22)) into the average power loss equation (equation (8.20)), one gets:

$$\left. \begin{aligned} K + K_2 \cdot (I_{eq})^2 &= \frac{1}{T_C}\sum_{i=1}^{n}\left(K + K_2 \cdot I_i^2\right) \cdot t_i = K + \frac{K_2}{T_C}\sum_{i=1}^{n}I_i^2 \cdot t_i \\ &\text{and} \\ K_2 \cdot (I_{eq})^2 &= \frac{K_2}{T_C}\sum_{i=1}^{n}I_i^2 \cdot t_i \end{aligned} \right\} \tag{8.23}$$

where
I_{eq} [A] is the equivalent current per cycle that would produce equal power loss in the motor, and
n is the total number of time-intervals in one cycle.

Finally, the equivalent current per cycle becomes:

$$I_{eq} = \sqrt{\frac{1}{T_C}\sum_{i=1}^{n}I_i^2 \cdot t_i} \ [\text{A}] \tag{8.24}$$

The equivalent current (equation (8.24)) allows for the selection of a motor from a catalogue. That is:

$$\left.\begin{array}{c} I_n^{(cat)} \geqslant I_{eq} \ [A] \\ \text{and} \\ V_n^{(cat)} = V_n \ [V] \\ n_n^{(cat)} = n_n \ [\text{rpm}] \end{array}\right\} \tag{8.25}$$

where
> $I_n^{(cat)}$ [A] is the catalogue (nominal) current of the motor,
> V_n [V] is the required terminal voltage,
> $V_n^{(cat)}$ [V] is the catalogue (nominal) voltage of the motor,
> n_n [rpm] is the required shaft speed, and
> $n_n^{(cat)}$ [rpm] is the catalogue (nominal) shaft speed.

As suggested before, a chosen motor based on the equivalent current I_{eq} (equation (8.24)) must also be assessed for ambient temperature (section 8.2) and for starting and maximum torques. If not adequate, a motor with higher rated power should be considered.

8.3.3.1 Piecewise linear approximation
The formula of the equivalent current (equation (8.24)) requires constant current at each time-interval. In practice, the current might vary with time during those intervals (solid line in figure 8.4). For hand calculation, a piecewise linear approximation of the current curve can be applied, which would simplify the rms computations.

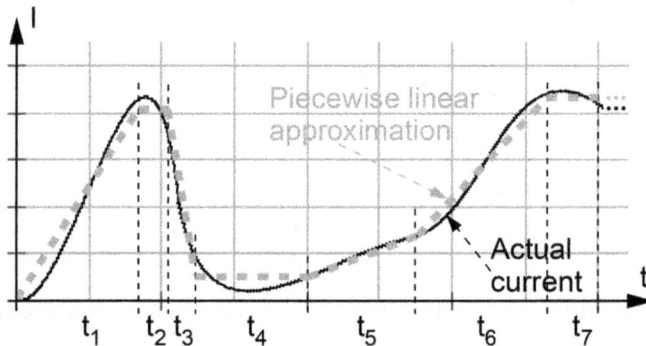

Figure 8.4. A load diagram of the motor current and its piecewise linear approximation.

The nonlinear nature of the current as a function of time can be approximated by using piecewise linear line segments (green dashed line in figure 8.4). At each time-interval, those segments display either a constant value (a horizontal straight line), or a linear curve starting or ending at a zero value (a triangular shape in figure 8.5

(a)), or a linear curve starting and ending at given values (a trapezoidal shape in figure 8.5(b)).

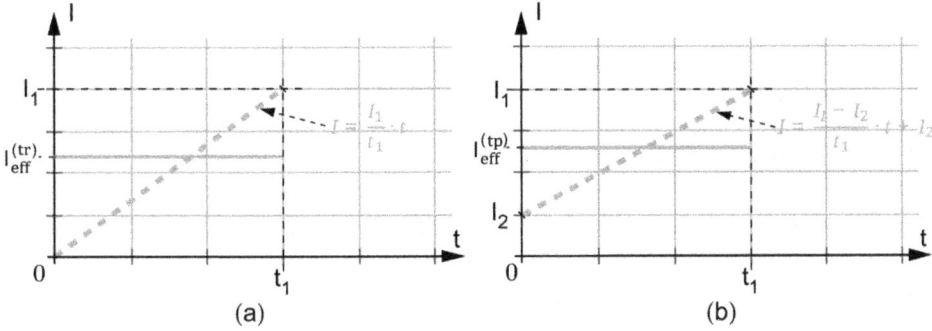

Figure 8.5. Linear line segments, (a) a triangular shape, and (b) a trapezoidal shape.

In figure 8.5, the dash lines present the piecewise linear current-segments of two randomly selected time-intervals—the first (figure 8.5(a)) has a triangular shape, and the second (figure 8.5(b)) has a trapezoidal shape. The bold lines present the effective (rms) value of those currents segments.

The effective value of the *triangular shape* (figure 8.5(a)) is:

$$I_{\text{eff}}^{(\text{tr})} = \sqrt{\frac{1}{t_1} \int_0^{t_1} \left(\frac{I_1}{t_1} \cdot t \right)^2 dt} = \frac{I_1}{\sqrt{3}} \ [\text{A}] \tag{8.26}$$

The effective value of the *trapezoidal shape* (figure 8.5(b)) is:

$$I_{\text{eff}}^{(\text{tp})} = \sqrt{\frac{1}{t_1} \int_0^{t_1} \left(\frac{I_1 - I_2}{t_1} \cdot t + I_2 \right)^2 dt} = \sqrt{\frac{I_1^2 + I_1 \cdot I_2 + I_2^2}{3}} \ [\text{A}] \tag{8.27}$$

The calculated effective (rms) value (equation (8.26) and (8.27)) can replace the ascending or the descending line segments of the piecewise linear approximation of the load diagram of figure 8.4.

8.3.4 Hand calculations—equivalent torque method

The equivalent torque method is based on the equivalent current equation that was developed above (equation (8.24)).

Assuming:

(a) constant magnetic flux in the airgap of the motor—examples are permanent magnet brushless DC motors, DC and synchronous motors having constant excitation current, and induction motors having constant terminal voltage —where the developed electromagnetic (EM) torque relates directly to the effective input current.

(b) the EM torque equals the shaft torque.

Then:

$$T_i \cong K' \cdot I_i \ [\text{Nm}] \tag{8.28}$$

where

T_i [Nm] is the torque at the motor shaft during time-interval i,
I_i [A] is the input current,
K' [Vs] is a coefficient that includes the constant magnetic flux, and
i denotes the time-interval number.

Substituting the current at each time-interval (equation (8.28)) into the equation for the effective current (equation (8.24)):

$$\left.\begin{aligned} \frac{T_{eq}}{K'} &\cong \sqrt{\frac{1}{T_C} \sum_{i=1}^{n} \left(\frac{T_i}{K'}\right)^2 \cdot t_i} \\ \text{and the equivalent } &\text{torque is:} \\ T_{eq} \cong \sqrt{\frac{1}{T_C} \sum_{i=1}^{n} T_i^2 \cdot t_i} &= \sqrt{\frac{T_1^2 t_1 + T_2^2 t_2 + \cdots + T_n^2 t_n}{t_1 + t_2 + \cdots + t_n}} \ [\text{Nm}] \end{aligned}\right\} \tag{8.29}$$

where

T_{eq} [Nm] is the equivalent **torque** of the motor,
T_i [Nm] is the **torque** at each time-interval,
T_C is the cycle **time**,
t_i is the interval **time**, and
n is the total number of time-intervals in one cycle.

The equivalent torque formula (equation (8.29)) allows for the selection of a motor from a catalogue. That is:

$$\left.\begin{aligned} P_n^{(cat)} &\geqslant T_{eq} \cdot \omega_n \ [\text{W}] \\ \text{and} \\ V_n^{(cat)} &= V_n \ [\text{V}] \\ n_n^{(cat)} &= \omega_n \frac{60}{2\pi} \ [\text{rpm}] \end{aligned}\right\} \tag{8.30}$$

where

$P_n^{(cat)}$ [W] is the catalogue (nominal) power of the motor,
ω_n [s^{-1}] is the required (rated) angular speed, and
$n_n^{(cat)}$ [rpm] is the catalogue shaft speed.

As suggested before, a selected motor based on equivalent torque T_{eq} (equation (8.29)) must be also assessed for ambient temperature (section 8.2) and for starting and maximum torques. If not adequate, a higher rated power of the motor must be considered.

8.3.5 Hand calculations—equivalent power method

The equivalent power method is based on the equivalent torque equation that was derived above (equation (8.29)). Assuming constant shaft speed throughout the cycle, then:

$$P_i \cong \omega_K \cdot T_i \ [W] \tag{8.31}$$

where

P_i [W] is the shaft power at time-interval i,
T_i [Nm] is the torque at the motor shaft at time-interval i,
ω_K [s^{-1}] is a constant angular speed of the motor, and
i denotes the time-interval number.

Note: The assumption of constant shaft speed prevents the use of this equivalent power method with a drive that includes starting and braking conditions.

Substituting the torque at each time-interval (equation (8.31)) into the equation for the effective torque (equation (8.29)):

$$\left.\begin{array}{c} \dfrac{P_{eq}}{\omega_K} \cong \sqrt{\dfrac{1}{T_C} \displaystyle\sum_{i=1}^{n}\left(\dfrac{P_i}{\omega_K}\right)^2 \cdot t_i} \\[2ex] \text{and the equivalent power is:} \\[2ex] P_{eq} \cong \sqrt{\dfrac{1}{T_C} \displaystyle\sum_{i=1}^{n} P_i^2 \cdot t_i} \ [W] \end{array}\right\} \tag{8.32}$$

The equivalent power formula, P_{eq} (equation (8.32)), allows for the selection of a motor from a catalogue. That is:

$$\left.\begin{array}{c} P_n^{(cat)} \geqslant P_{eq} \ [W] \\[1ex] \text{and} \\[1ex] V_n^{(cat)} = V_n \ [V] \\[1ex] n_n^{(cat)} = \omega_n \dfrac{60}{2\pi} \ [\text{rpm}] \end{array}\right\} \tag{8.33}$$

where $P_n^{(cat)}$ [W] is the catalogue (nominal) power of the motor.

As suggested before, a selected motor based on the equivalent power P_{eq} (equation (8.32)) must also be assessed for ambient temperature (section 8.2) and for starting and maximum torques. If not adequate, a motor with higher rated power should be considered.

8.4 Motor selection at short-time-operating mode

The duty time at short-time-operating mode is in the order of the thermal time constant of the motor (section 7.5.2). The motor temperature does not reach a stable value during that duty time, and it has enough time to reach ambient temperature at downtime.

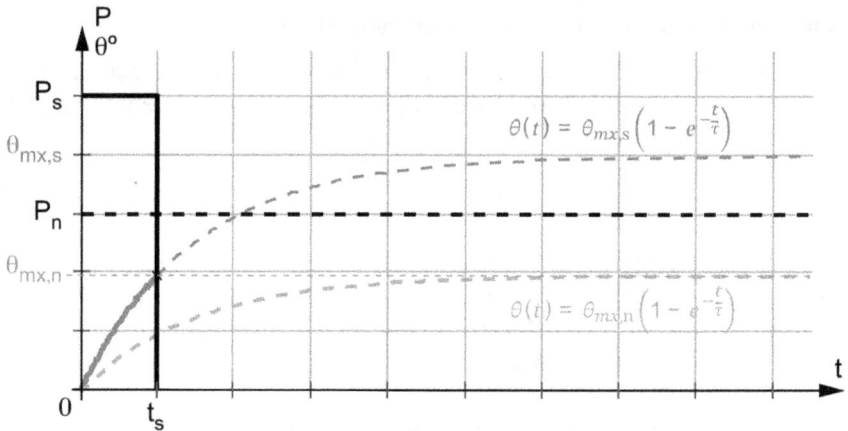

Figure 8.6. Motor heating during short-time-operating mode.

Continuous operation at rated power P_n (dashed line in figure 8.6) would bring the motor temperature to its rated (allowed) maximum value:

$$\left.\begin{array}{r} \theta(t) = \theta_{mx,n}\left(1 - e^{-\frac{t}{\tau}}\right) \ [°C] \\[2mm] \theta_{mx,n} = \dfrac{\Delta P_n}{A} \ [°C] \end{array}\right\} \tag{8.34}$$

where
    ΔP_n [W] is the motor power loss when it is operating at rated power
    $\theta_{mx,n}$ [°C] is the rated (allowed) maximum temperature (see equation (7.9)), and
    τ [sec] is the thermal time constant (see equation (7.11)).

During a short-time-interval t_s, the shaft power can be increased (P_s in figure 8.6) as long as the maximum temperature of the motor would not rise above its rated (allowed) maximum temperature:

$$\left.\begin{array}{r} \theta_{mx,n} = \theta_{mx,s}\left(1 - e^{-\frac{t_s}{\tau}}\right) \ [°C] \\[2mm] \theta_{mx,s} = \dfrac{\Delta P_s}{A} \ [°C] \end{array}\right\} \tag{8.35}$$

where
    $\theta_{mx,s}$ [°C] is the maximum temperature that the motor can reach when operating continuously at a shaft power of P_s,
    P_s [W] is the short-time operating power,
    ΔP_s [W] is the power loss when the motor operates at P_s, and
    t_s is the short-time-operating interval (duty time).

Substituting the rated maximum temperature (equation (8.34)) into equation (8.35):

$$\left.\begin{aligned}
\frac{\Delta P_n}{A} &= \frac{\Delta P_s}{A}\left(1 - e^{-\frac{t_s}{\tau}}\right) \\
\text{and} \\
P_H = \frac{\theta_{mx,s}}{\theta_{mx,n}} &= \frac{\Delta P_s}{\Delta P_n} = \frac{1}{1 - e^{-\frac{t_s}{\tau}}}
\end{aligned}\right\} \tag{8.36}$$

where $P_H = f(t_s)$ is defined as the overload heating factor.

The normalized shaft power can be obtained by substituting the motor power loss (equation (7.6)) into the overload heating factor (equation (8.36)):

$$\left.\begin{aligned}
P_H = \frac{\Delta P_s}{\Delta P_n} &= \frac{P_{cu,n}(a + x^2)}{P_{cu,n}(a + 1^2)} \\
\text{and} \\
x &= \sqrt{P_H(a + 1) - a}
\end{aligned}\right\} \tag{8.37}$$

where x is the normalized shaft power of the motor (equation (7.5)).

Finally, the allowed short-time power would be:

$$P_s = x \cdot P_n \ [\text{W}] \tag{8.38}$$

Note: In a short-time operating mode, the startup process plays a significant role in the motor heating. When a self-cooled motor is used where ventilation is provided by an internal fan mounted on the shaft, the heat transfer coefficient is affected by the shaft speed (addressed in a succeeding section 8.5.2).

Example 8.2 The nameplate of a DC motor states: $V_n = 220$ V, $P_n = 30$ kW, $I_n = 157$ A, $n_n = 980$ rpm. The motor is designed for a continuous operation mode. On two occasions, the motor was operated at short-time processes:

(a) $t_{s1} = 60$ min, $P_1 = 32.5$ kW, and $I_1 = 170$ A
(b) $t_{s2} = 30$ min, $P_2 = 35$ kW, and $I_2 = 183$ A

Calculate the thermal time constant in both cases.

Solution

At rated load, the power lose is:

$$\Delta P_n = P_{in} - P_n \cong 220 \cdot 157 - 30 \cdot 10^3 = 4.54 \ [\text{kW}]$$

(a) At $P_1 = 32.5$ kW, the power lose is:

$$\Delta P_1 = P_{in} - P_1 \cong 220 \cdot 170 - 32.5 \cdot 10^3 = 4.9 \ [\text{kW}]$$

The overload heating factor is (equation (8.36)):

$$P_{H1} = \frac{\Delta P_1}{\Delta P_n} = \frac{4.9}{4.54} = 1.08$$

The thermal time constant is:

$$\tau_1 = \frac{t_{s1}}{\ln \dfrac{P_{H1}}{P_{H1} - 1}} = \frac{60}{\ln \dfrac{1.08}{1.08 - 1}} = 23 \ [\text{min}]$$

(b) At $P_2 = 35$ kW, the power lose is:

$$\Delta P_2 = P_{in} - P_2 \cong 220 \cdot 183 - 35 \cdot 10^3 = 5.26 \ [\text{kW}]$$

The overload heating factor is:

$$P_{H2} = \frac{5.26}{4.54} = 1.16$$

And the thermal time constant is:

$$\tau_2 = \frac{60}{\ln \dfrac{1.16}{1.16 - 1}} = 15 \ [\text{min}]$$

In practice, use of the overload heating factor (equation (8.36)) requires information about the thermal time constant, which is not provided in the motor catalogue. Moreover, a general-purpose (continuous operation mode) motor is not recommended for an overload short-time operation because its internal conductors might overheat which would cause permanent damage to the insulating materials. For that reason, a dedicated motor design is required and the industrial market does offer special catalogues for short-time-operation motors.

8.4.1 Catalogue selection

Short-time-operating motors are designed to distribute the excessive heat generated in the conductors evenly around the machine. The catalogue of those motors suggests, among other parameters, the rated power at a specified time-interval. For instance, a nameplate of 15 kW and 30 min indicates that after operating at 15 kW for 30 min, the motor must be shut down and remains at rest until it reaches ambient temperature.

To adopt a motor having non-standard working conditions from a catalogue for short-time-operating motors, the selection must ensure that the rated maximum temperature would not be compromised.

Consider an engineering design that requires a short-time-operating interval t_s of a motor at a peak power P_s (figure 8.7). But the standard catalogue offers a motor with a short-time-interval $t_n^{(cat)}$ at a peak power $P_n^{(cat)}$. In both cases, at the end of the short-time-interval, the temperature must reach the rated value $\theta_{mx,n}$. Using the temperature heating formula (equation (7.11) where $\theta_0 = 0°$):

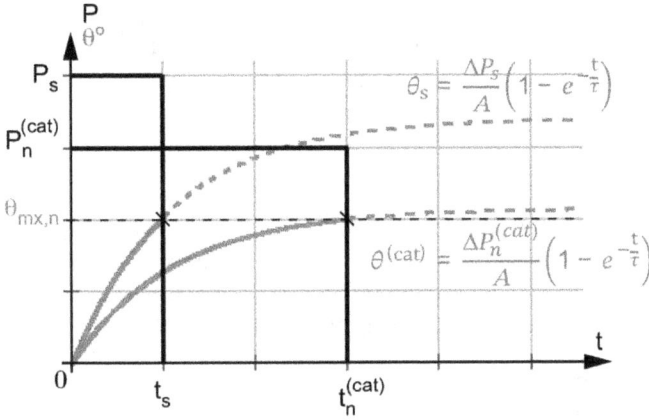

Figure 8.7. Motor heating at short-time operation Short-time power demand compared to catalogue standard.

$$\left.\begin{array}{r}
\theta_{mx,n} = \dfrac{\Delta P_n^{(cat)}}{A}\left(1 - e^{-\frac{t_s^{(cat)}}{\tau}}\right) = \dfrac{\Delta P_s}{A}\left(1 - e^{-\frac{t_s}{\tau}}\right) \\[2mm]
\text{using the exponential approximation (equation (8.19)):} \\[2mm]
\dfrac{\Delta P_n^{(cat)}}{A} \cdot \dfrac{t_s^{(cat)}}{\tau} \cong \dfrac{\Delta P_s}{A} \cdot \dfrac{t_s}{\tau} \\[2mm]
\text{and the power loss ratio is: } \dfrac{\Delta P_s}{\Delta P_n^{(cat)}} \cong \dfrac{t_s^{(cat)}}{t_s}
\end{array}\right\} \qquad (8.39)$$

where
$\Delta P_n^{(cat)}$ [W] is the power loss at rated (catalogue) power $P_n^{(cat)}$,
$t_s^{(cat)}$ is the catalogue short-time-interval,
ΔP_s [W] is the power loss at the required short-time power P_s, and
t_s is the required short-time-interval.

Substituting the power loss formula (equation (7.6)) into the power loss ratio above (equation (8.39)):

$$\left.\begin{array}{r}
\dfrac{\Delta P_s}{\Delta P_n^{(cat)}} = \dfrac{P_{cu,\,n}(a + x^2)}{P_{cu,\,n}(a + 1^2)} \cong \dfrac{t_s^{(cat)}}{t_s} \\[2mm]
\text{and the normalized shaft power becomes:} \\[2mm]
x \cong \sqrt{(a + 1)\dfrac{t_s^{(cat)}}{t_s} - a}
\end{array}\right\} \qquad (8.40)$$

Finally, the selected motor from the short-time-operation catalogue should be:

$$P_n^{(cat)} \geqslant \dfrac{P_s}{x} \;\; [\text{W}] \qquad\qquad (8.41)$$

where
P_s [W] is the required operating power at the short-time of t_s, and
x is the normalized shaft power (see equation (7.5)).
Note: The selected motor should be assessed for ambient conditions (section 8.2).

8.5 Motor selection at intermittent cycle operation mode

The duty time at intermittent cycle operating mode, also called intermittent periodic duty, is in the order of the thermal time constant of the motor (section 7.5.3). During the duty time, the motor temperature does not reach a stable value. During the downtime, the motor temperature cannot reach the ambient value.

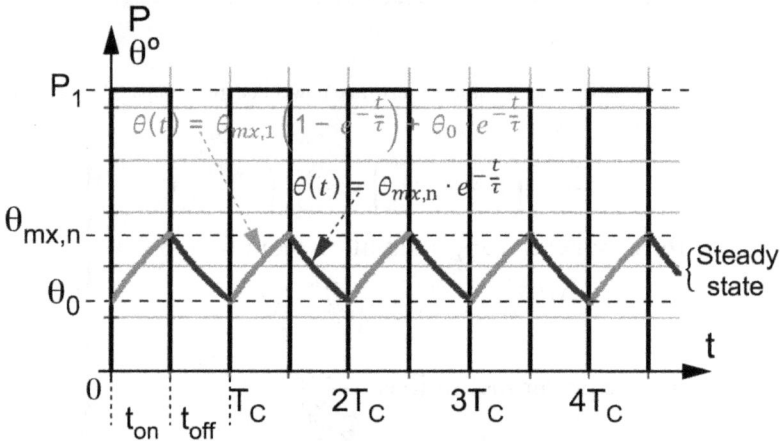

Figure 8.8. Thermal steady-state condition during intermittent cycle operation mode.

Practical thermal steady-state operation (figure 8.8) suggests that during each cycle, the motor temperature would be allowed to reach the maximum rated value. Using equation (7.17), that rated value is:

$$\theta_{mx,n} = \theta_{mx,1} \left(1 - e^{-\frac{t_{on}}{\tau}} \right) + \theta_0 \cdot e^{-\frac{t_{on}}{\tau}} \quad [°C] \tag{8.42}$$

where
$\theta_{mx,n}$ [°C] is the rated (allowed) maximum temperature,
$\theta_{mx,1}$ [°C] is the maximum temperature during continuous operation at P_1 [W],
θ_0 [°C] is the initial temperature at each operating cycle,
t_{on} is the duty time at each operating cycle, and
τ is the thermal time constant of the motor.

Assuming the same thermal time constant during downtime, the motor temperature would be (see equation (7.18)):

$$\theta_0 = \theta_{mx,n} \cdot e^{-\frac{t_{off}}{\tau}} \quad [°C] \tag{8.43}$$

where t_{off} is the downtime at each operating cycle.

Substituting θ_0 (equation (8.43)) into equation (8.42), one gets:

$$\left. \begin{array}{c} \theta_{mx,n} = \theta_{mx,1} \left(1 - e^{-\frac{t_{on}}{\tau}} \right) + \theta_{mx,n} \cdot e^{-\frac{T_C}{\tau}} \; [°C] \\[2mm] \text{or} \\[2mm] \theta_{mx,n} \left(1 - e^{-\frac{T_C}{\tau}} \right) = \theta_{mx,1} \left(1 - e^{-\frac{t_{on}}{\tau}} \right) \end{array} \right\} \qquad (8.44)$$

where $T_C = t_{on} + t_{off}$ is the cycle time

The overload heating factor (equation (8.36)) becomes:

$$P_H = \frac{\theta_{mx,1}}{\theta_{mx,n}} = \frac{1 - e^{-\frac{T_C}{\tau}}}{1 - e^{-\frac{t_{on}}{\tau}}} = \frac{1 - e^{-\frac{t_{on}}{\tau} \frac{1}{\varepsilon}}}{1 - e^{-\frac{t_{on}}{\tau}}} \qquad (8.45)$$

where ε is the duty cycle (equation (7.16)).

The behavior of the overload heating factor as a function of the duty cycle, $P_H = f(\varepsilon)$, would be [3]:

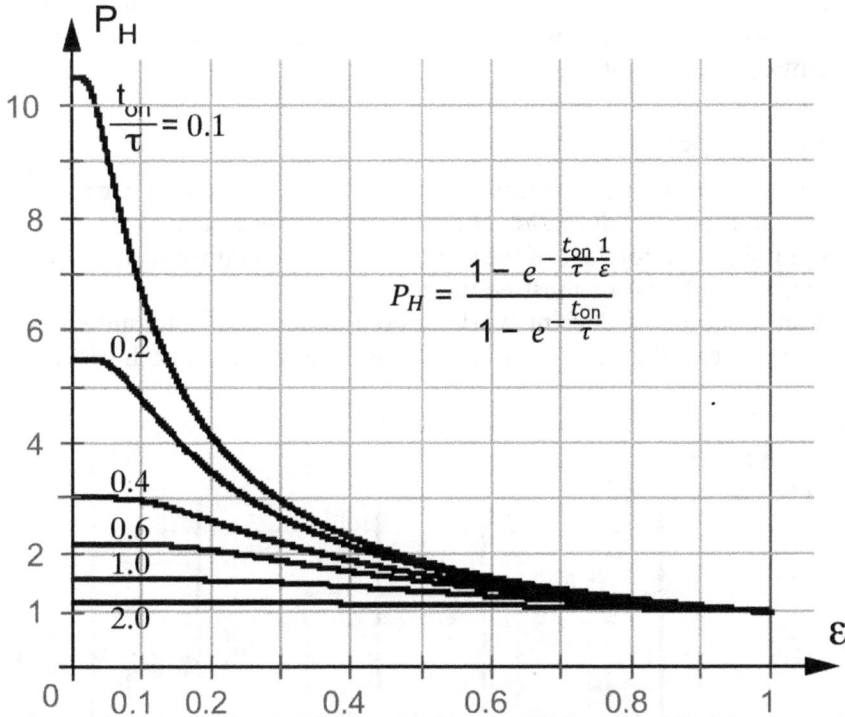

Figure 8.9. The overload heating factor as a function of the duty cycle.

As displays in figure 8.9, at small duty cycles, the value of the overload heating factor at intermittent operation mode (equation (8.45)) is about the same as for short-time operation (equation (8.36)):

$$P_H = \frac{1 - e^{-\frac{t_{on}}{\tau} \frac{1}{\varepsilon}}}{1 - e^{-\frac{t_{on}}{\tau}}} \overset{\varepsilon \ll 1}{\Longrightarrow} \frac{1}{1 - e^{-\frac{t_s}{\tau}}} \qquad (8.46)$$

As the duty cycle approaches unity, the overload heating factor also reaches unity, which implies continuous operating mode:

$$P_H = \frac{1 - e^{-\frac{t_{on}}{T}\frac{1}{\varepsilon}}}{1 - e^{-\frac{t_{on}}{T}}} \xrightarrow{\varepsilon \lesssim 1} 1 \tag{8.47}$$

The above (figure 8.9, equations (8.46) and (8.47)) suggest practical boundaries for short-time, intermittent, and continuous operating modes:

$$\left.\begin{array}{l} 0 < \varepsilon \leqslant 0.1 \Longrightarrow \text{Short-time operating mode} \\ 0.1 < \varepsilon \leqslant 0.6 \Longrightarrow \text{Intermittent operating mode} \\ 0.6 < \varepsilon \leqslant 1 \Longrightarrow \text{Continuous operating mode} \end{array}\right\} \tag{8.48}$$

In a cyclic mode of operation, when the duty cycle ε is very small, $0 < \varepsilon \leqslant 0.1$, motor operation is regarded as short-time mode (equation (8.48)). When the duty cycle is between $0.1 < \varepsilon \leqslant 0.6$, motor operation is regarded as intermittent mode. When the duty cycle is between $0.6 < \varepsilon \leqslant 1$, a continuous operation mode should be considered.

As with short-time-operating motors, a dedicated motor design is available for intermittent operation motors.

8.5.1 Catalogue selection

A catalogue for intermittent operating motors suggests, among other parameters, a rated power at a specified duty cycle. For instance, a motor nameplate of 15 kW and $S3\,25\%$ [4] implies that for 25% of the cycle, the motor can operate at 15 kW, and for 75% of the cycle, the motor must be turned off.

To adopt a motor of deviant working conditions from intermittent operation catalogue standards, the selection must ensure equal losses in both cases.

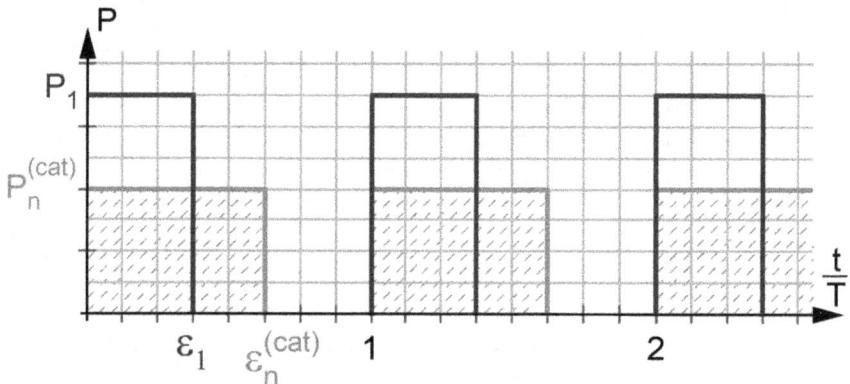

Figure 8.10. Intermittent cycle operation mode. Duty cycle demand compared with a motor catalogue standard.

Consider an engineering design that requires a peak power P_1 at a duty cycle ε_1 (figure 8.10). The standard catalogue offers a peak power $P_n^{(cat)}$ at a duty cycle $\varepsilon_n^{(cat)}$.

In both cases, the total power loss per cycle must be the same. Using the power loss equation (equation (7.6)):

$$\Delta P = P_{cu,\,n}(a + 1^2) \cdot \varepsilon_n^{(cat)} = P_{cu,\,n}(a + x^2) \cdot \varepsilon_1 \;\; [\text{W}]$$

and the normalized shaft power would be:

$$x = \sqrt{(a + 1)\frac{\varepsilon_n^{(cat)}}{\varepsilon_1} - a}$$

$$(8.49)$$

Finally, the selected motor from the intermittent operation catalogue should be (equation (8.7)):

$$P_n^{(cat)} \geqslant \frac{P_1}{x} \;\; [\text{W}] \tag{8.50}$$

where P_1 [W] is the required peak power at a duty cycle ε_1.

Note: The selected motor should be assessed for ambient conditions (section 8.2).

Example 8.3 An induction motor drives a mechanism. The motor heat dissipation is provided externally by a forced cooling unit. At full load, the motor copper losses are 60% of its total power losses. The load diagram at the motor shaft is given in figure E8.3, where the cycle time is $T_C = 16$ min. The rated speed of the motor is $n_n = 1750$ rpm.

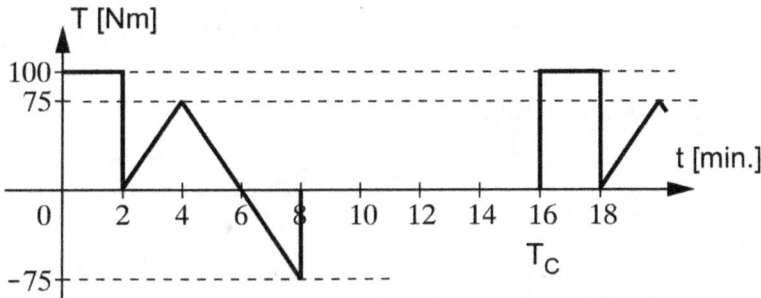

Figure E8.3. Load diagram at the shaft of the induction motor.

A catalogue for intermittent operating motors offers the following data:

$$\begin{cases} \text{Motor } \#1: \quad P_{n1}^{(cat)} = 12 \text{ kW at } \varepsilon_{n1}^{(cat)} = 0.4 \\ \text{Motor } \#2: \quad P_{n2}^{(cat)} = 11 \text{ kW at } \varepsilon_{n2}^{(cat)} = 0.5 \\ \text{Motor } \#3: \quad P_{n3}^{(cat)} = 10 \text{ kW at } \varepsilon_{n3}^{(cat)} = 0.6 \end{cases}$$

Select the appropriate motor for the given load diagram.

Solution
The duty cycle of the drive is:

$$\varepsilon_1 = \frac{8}{16} = 0.5$$

Suggesting that a catalogue for intermittent operating motors should be used (equation (8.48)).
The motor loss coefficient is (equation (7.6)):

$$a = \frac{0.4}{0.6} = 0.67$$

During the time-interval $2 \leqslant t < 4$ min, the equivalent torque is (equation (8.26)):

$$T_{2,4} = \frac{75}{\sqrt{3}} = 43.3 \ [\text{Nm}]$$

During the time-interval $4 \leqslant t < 8$ min, the equivalent torque is (equation (8.27)):

$$T_{4,8} = \sqrt{\frac{75^2 + 75(-75) + (-75)^2}{3}} = 43.3 \ [\text{Nm}]$$

The combined equivalent torque during the duty cycle, that is, during the time-interval $0 \leqslant t < 8$ min, is (equation (8.29)):

$$T_{eq} \equiv T_{0,8} = \sqrt{\frac{100^2 \cdot 2 + 43.3^2 \cdot 2 + 43.3^2 \cdot 4}{8}} = 62.5 \ [\text{Nm}]$$

The corresponding power during the duty cycle is:

$$P_{0,8} = T_{0,8} \cdot \omega_n = 62.5\frac{2\pi \cdot 1750}{60} = 11.45 \ [\text{kW}]$$

Consider the three catalogue motors given in the problem, the allowed intermittent power of each at the required duty cycle of $\varepsilon_1 = 0.5$ is (equation (8.50)):

$$\left\{ \begin{array}{l} \text{Motor \#1:} \ \ P_1 = x_1 \cdot P_{n1}^{(cat)} = \sqrt{(0.67 + 1)\dfrac{0.4}{0.5} - 0.67} \cdot 12 = 0.816 \cdot 12 = 9.79 \ [\text{kW}] \\[2mm] \text{Motor \#2:} \ \ P_2 = x_2 \cdot P_{n2}^{(cat)} = \sqrt{(0.67 + 1)\dfrac{0.5}{0.5} - 0.67} \cdot 11 = 1.0 \cdot 11 = 11 \ [\text{kW}] \\[2mm] \text{Motor \#3:} \ \ P_3 = x_3 \cdot P_{n3}^{(cat)} = \sqrt{(0.67 + 1)\dfrac{0.6}{0.5} - 0.67} \cdot 10 = 1.155 \cdot 10 = 11.55 \ [\text{kW}] \end{array} \right\}$$

Finally, motor #3 should be selected from the catalogue because $P_3 > P_{0,8}$. See figure E8.3.1.

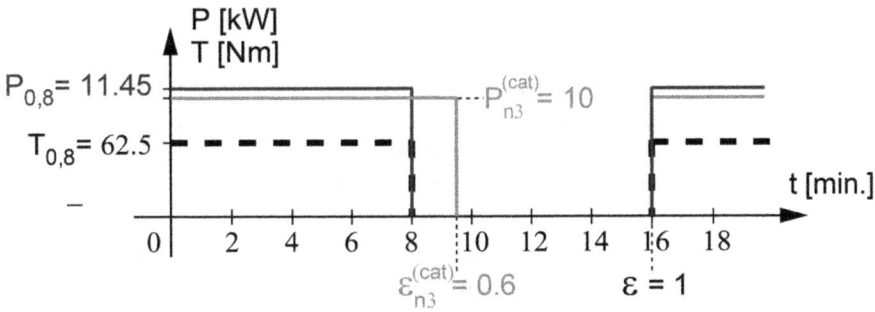

Figure E8.3.1. Duty cycle demand compared with the motor catalogue standard.

8.5.2 Thermal considerations of self-ventilated motors

A self-ventilated, or self-cooled motor means that ventilation is provided by an internal fan mounted on the shaft and rotates with it. For that reason, the motor heat transfer conditions differ during its startup, work, brake and rest periods. Consequently, the thermal time constant (equation (7.11)) varies too. That would affect the selection of the rated (nominal) power of the motor.

Consider a basic load diagram that includes four time-intervals: startup, working, braking, and rest periods of the driving motor.

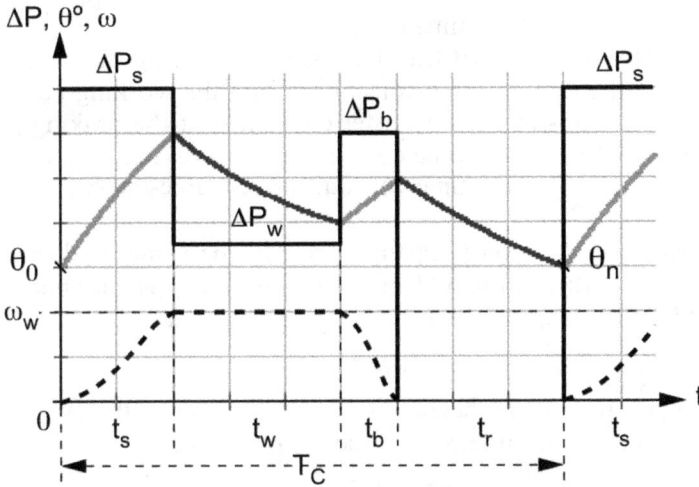

Figure 8.11. Load diagram that includes startup, working, braking and rest periods.

The load diagram (figure 8.11) shows the behavior of the power loss ΔP during each time-interval (bold line), the rise and fall of the motor temperature $\theta°$ (red and blue lines), and the shaft speed ω (dashed line) as a function of time. The cycle time itself is:

$$T_C = t_s + t_w + t_b + t_r \tag{8.51}$$

where

T_C is the cycle time,

t_s is the time-interval during motor startup (acceleration),

t_w is the time-interval during working conditions,

t_b is the time-interval during braking mode (deceleration), and

t_r is the time-interval where the motor is at rest (standstill).

At the startup period t_s, the shaft speed increases, $0 \leqslant \omega < \omega_w$. During the working period t_w, the shaft speed remains constant at ω_w. During the braking period t_b, the shaft speed decreases, $\omega_w > \omega \geqslant 0$. And during the rest period t_r, the motor is turned off.

Given a load diagram of the power at the motor shaft, the power loss ΔP at each time-interval can be calculated (equation (8.9)). Following the discussion in section 8.3.2, the motor temperature at the end of the cycle can also be calculated:

$$
\left.
\begin{aligned}
\theta_n = {} & \frac{\Delta P_s}{A_s}\left(1 - e^{-\frac{t_s}{\tau_s}}\right)e^{-\left(\frac{t_w}{\tau} + \frac{t_b}{\tau_b} + \frac{t_r}{\tau_r}\right)} + \frac{\Delta P_w}{A}\left(1 - e^{-\frac{t_w}{\tau}}\right)e^{-\left(\frac{t_b}{\tau_b} + \frac{t_r}{\tau_r}\right)} \\
& + \frac{\Delta P_b}{A_b}\left(1 - e^{-\frac{t_b}{\tau_b}}\right)e^{-\left(\frac{t_r}{\tau_r}\right)} + \theta_0 \cdot e^{-\left(\frac{t_s}{\tau_s} + \frac{t_w}{\tau} + \frac{t_b}{\tau_b} + \frac{t_r}{\tau_r}\right)}
\end{aligned}
\right\}
\tag{8.52}
$$

where

θ_n [°C] is the motor temperature at the end of the cycle,

θ_0 [°C] is the motor temperature at the beginning of the cycle,

A_s [joule s^{-1} °C^{-1}] is the heat transfer coefficient at startup,

τ_s [s or min] is the thermal time constant at startup,

A [joule s^{-1} °C^{-1}] is the heat transfer coefficient during the working period,

τ [s or min] is the thermal time constant during the working period,

A_b [joule s^{-1} °C^{-1}] is the heat transfer coefficient at the braking period,

τ_b [s or min] is the thermal time constant at the braking period, and

τ_r [s or min] is the thermal time constant at rest time-interval.

The heat transfer during motor startup is nearly the same as during its braking mode. A practical deduction would be to assume that the thermal time constant (equation (7.11)) is the same in those two operating modes:

$$
\tau_s \cong \tau_b
\tag{8.53}
$$

The motor temperature at the beginning and at the end of the cycle are the same, $\theta_0 = \theta_n$. As a result, equation (8.52) can be presented as:

$$
\left.
\begin{aligned}
\theta_n\left[1 - e^{-\left(\frac{t_s + t_b}{\tau_s} + \frac{t_w}{\tau} + \frac{t_r}{\tau_r}\right)}\right] = {} & \frac{\Delta P_s}{A_s}\left(1 - e^{-\frac{t_s}{\tau_s}}\right)e^{-\left(\frac{t_w}{\tau} + \frac{t_b}{\tau_b} + \frac{t_r}{\tau_r}\right)} \\
& + \frac{\Delta P_w}{A}\left(1 - e^{-\frac{t_w}{\tau}}\right)e^{-\left(\frac{t_b}{\tau_b} + \frac{t_r}{\tau_r}\right)} + \frac{\Delta P_b}{A_b}\left(1 - e^{-\frac{t_b}{\tau_b}}\right)e^{-\left(\frac{t_r}{\tau_r}\right)}
\end{aligned}
\right\}
\tag{8.54}
$$

As addressed before (equation (8.17)), the left term of equation (8.54) suggests that when the motor operates continuously at an average power loss ΔP_{avg}, its temperature would reach a steady-state peak value of θ_n:

$$\theta_n = \frac{\Delta P_{\text{avg}}}{A} \ [°C] \tag{8.55}$$

Applying the exponential approximation (equation (8.19)) to the thermal equation above (equation (8.54)), one gets:

$$\left.\begin{array}{c} \dfrac{\Delta P_{\text{avg}}}{A}\left(\dfrac{t_s + t_b}{\tau_s} + \dfrac{t_w}{\tau} + \dfrac{t_r}{\tau_r}\right) \cong \dfrac{\Delta P_s}{A_s} \cdot \dfrac{t_s}{\tau_s} + \dfrac{\Delta P_w}{A} \cdot \dfrac{t_w}{\tau} + \dfrac{\Delta P_b}{A_b} \cdot \dfrac{t_b}{\tau_b} \\[2mm] \text{multiplying both sides of the equation above by } (A \cdot \tau), \text{ one gets:} \\[2mm] \Delta P_{\text{avg}}\left[(t_s + t_b)\dfrac{\tau}{\tau_s} + t_w + t_r\dfrac{\tau}{\tau_r}\right] \cong \Delta P_s \cdot t_s\dfrac{A}{A_s} \cdot \dfrac{\tau}{\tau_s} + \Delta P_w \cdot t_w + \Delta P_b \cdot t_b\dfrac{A}{A_s} \cdot \dfrac{\tau}{\tau_s} \end{array}\right\} \tag{8.56}$$

The term τ/τ_s, the thermal time constant during work mode over the thermal time constant during startup mode suggests:

$$\frac{\tau}{\tau_s} = \frac{\dfrac{C}{A}}{\dfrac{C}{A_s}} = \frac{A_s}{A} \tag{8.57}$$

substituting the above ratio (equation (8.57)) into equation (8.56) provides the average power loss of the motor when it operates continuously:

$$\left.\begin{array}{c} \Delta P_{\text{avg}} = \dfrac{\Delta P_s \cdot t_s + \Delta P_w \cdot t_w + \Delta P_b \cdot t_b}{(t_s + t_b) \cdot \alpha + t_w + t_r \cdot \beta} \\[3mm] \text{where: the start/brake cooling coefficient } \alpha = \dfrac{\tau}{\tau_s} \\[3mm] \text{and, the rest cooling coefficient } \beta = \dfrac{\tau}{\tau_r} \end{array}\right\} \tag{8.58}$$

With a self-cooled motor, the ventilation improves as the shaft speed increases, which leads to varying thermal time constants; that is: $\tau < \tau_s < \tau_r$. Suggested empirical values are [3]:

$$\left.\begin{array}{l} \text{DC motors: } \alpha \cong 0.75 \text{ and } \beta \cong 0.5 \\ \text{AC motors: } \alpha \cong 0.5 \text{ to } 0.6 \text{ and } \beta \cong 0.25 \end{array}\right\} \tag{8.59}$$

The required shaft power itself can be calculated using equation (8.21).

Note: Following the analysis presented in sections 8.3.3–8.3.5, the inclusion of the start/brake cooling coefficient α and the rest cooling coefficient β (equation (8.58)) in the average power loss formula (equation (8.58)) can be applied to the equivalent current (equation (8.24)), to the equivalent torque (equation (8.29)), and to the equivalent power (equation (8.32)) formulas.

8.6 Numerical example

An industrial rolling machine uses a solid drum to crush gravel and rocks (figure E8.6). The self-ventilated induction motor and the gearbox are mounted on a cart that moves on two rails together with the rotating drum. The entire electromechanical drive moves back and forth along the rails in a cyclic process that includes a rest time to clear and reload the gravel.

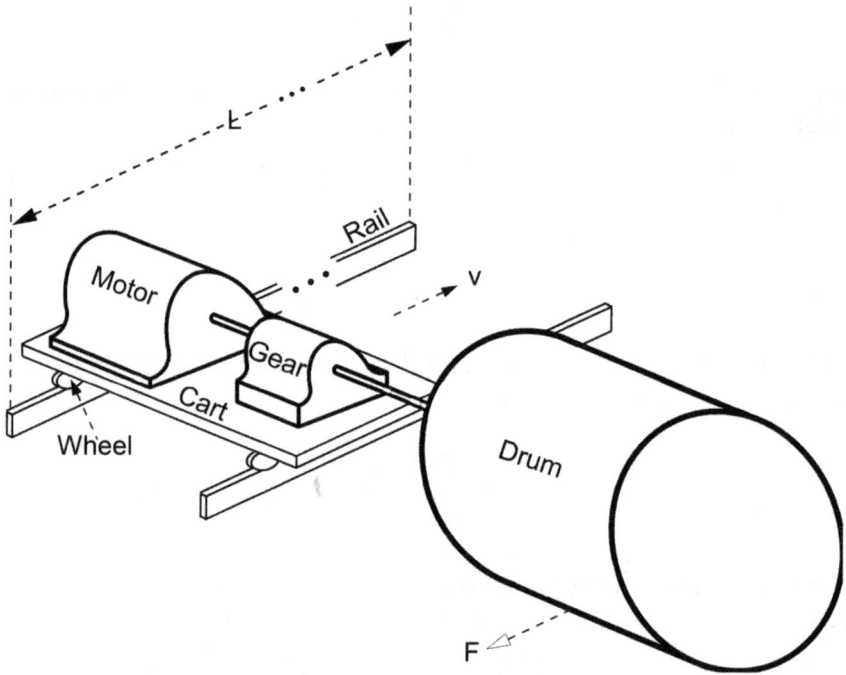

Figure E8.6. An industrial rolling machine.

Given:

(a) The track length is $L = 1208$ m.

(b) The rated linear velocity of the drive is: $v = 16$ m s^{-1}.

(c) The gearbox ratio is: $i = 4.5$.

(d) The gearbox efficiency is: $\eta_g = 0.95$.

(e) The flywheel moment of the gearbox is: $(GD^2)_g = 26.4$ Nm^2.

(f) The friction of the rolling drum with the gravels is: $F = 116$ kg$_f$.

(g) The diameter of the drum is: $D_D = 2$ m.

(h) The flywheel moment of the drum is: $(GD^2)_D = 6.8$ kNm^2.

(i) The weight of the cart, motor, gearbox, and wheels is $G_C = 136$ kg$_f$.

(j) The diameter of each cart-wheel is: $D_w = 0.2$ m.

(k) The flywheel moment of each wheel, out of 4, is: $(GD^2)_w = 3.2$ Nm^2.

(l) The acceleration of the drive during startup is: $a_s = 1$ m s^{-2}.

(m) The deceleration of the drive during brake is: $a_b = 1.6$ m s^{-2}.

(n) The cycle time is: $T_C = 145$ s.

Neglect the flywheel moment of the cart's wheels and their friction with the rails, and address the following:

(a) Build the load diagram of the drive; that is, motor-toque versus time.

(b) Select the rated (nominal) power of the self-ventilated induction motor.

Solution

The angular speed of the cart's wheels is:

$$\omega_w = \frac{v}{\frac{D_w}{2}} = \frac{16}{\frac{0.2}{1}} = 160 \ [s^{-1}]$$

The angular speed of the drum is:

$$\omega_D = \frac{v}{\frac{D_D}{2}} = \frac{16}{\frac{2}{2}} = 16 \ [s^{-1}]$$

The rated speed of the motor is:

$$\omega_n = i \cdot \omega = 4.5 \cdot 16 = 72 \ [s^{-1}] \quad \{n_n = 687.6 \ \text{rpm}\}$$

The resistive load torque referred to the motor shaft is:

$$T_r = F \frac{D_D}{2} \cdot \frac{1}{i \cdot \eta_g} = 116 \cdot 9.81 \frac{2}{2} \cdot \frac{1}{4.5 \cdot 0.95} = 266.2 \ [\text{Nm}]$$

Motor power approximation

To obtain the flywheel moment (equation (1.10)) of the motor itself, one needs first to estimate its rated power, and then to select one from a motor catalogue.

Using a coefficient of safety of 1.2, the estimated shaft power would be:

$$P_{est} = 1.2 \cdot T_r \cdot \omega_n = 1.2 \cdot 266.2 \cdot 72 = 23 \ [\text{kW}]$$

From a motor catalogue, the flywheel moment is $(GD^2)_m = 51.6 \ [\text{Nm}^2]$

(a) *Calculation of the operating time-intervals*

The acceleration time is:

$$t_s = \frac{v}{a_s} = \frac{16}{1} = 16 \ [s]$$

The distance during motor startup (during acceleration) is:

$$L_a = \frac{1}{2} \cdot a_s \cdot t_s^2 = \frac{1}{2} \cdot 1 \cdot 16^2 = 128 \ [m]$$

The deceleration time is:

$$t_b = \frac{v}{a_b} = \frac{16}{1.6} = 10 \ [s]$$

The distance during braking mode (during deceleration) is:

$$L_b = v \cdot t_b - \frac{1}{2} \cdot a_b \cdot t_b^2 = 16 \cdot 10 - \frac{1}{2} \cdot 1.6 \cdot 10^2 = 80 \ [m]$$

The distance during work mode (during rated speed of the motor) is:

$$L_w = L - (L_a + L_b) = 1208 - (128 + 80) = 1000 \ [m]$$

The time during rated speed is:

$$t_w = \frac{L_w}{v} = \frac{1000}{16} = 62.5 \ [s]$$

The rest time is:

$$t_r = T_C - (t_s + t_w + t_b) = 145 - (16 + 62.5 + 10) = 56.5 \ [\text{s}]$$

The duty cycle is:

$$\varepsilon = \frac{t_s + t_w + t_b}{T_C} = \frac{16 + 62.5 + 10}{145} = 0.61$$

As suggested above (equation (8.48)), a motor for continuous operation should be selected.

Calculation of the referred flywheel moment referred (see chapter 2):

$$
\left\{
\begin{aligned}
&= (GD^2)_{shaft} \\
&= (GD^2)_m + (GD^2)_g + (GD^2)_D \left(\frac{\omega_D}{\omega_m}\right)^2 + 4 \cdot (GD^2)_w \left(\frac{\omega_w}{\omega_m}\right)^2 + 365 \cdot G_C \left(\frac{v}{n_n}\right)^2 \\
&= 51.6 + 26.4 + 6800 \left(\frac{16}{72}\right)^2 + 4 \cdot 3.2 \left(\frac{160}{72}\right)^2 + 365 \cdot 136 \cdot 9.81 \left(\frac{16}{687.6}\right)^2 \\
&= 740.7 \ [\text{Nm}^2]
\end{aligned}
\right.
$$

Calculation of the motor torque during startup and braking modes:

The required torque at the motor shaft during the startup mode is (equation (1.12)):

$$T_{m,s} = T_r + \frac{(GD^2)_{shaft}}{375} \frac{dn}{dt} = 266.2 + \frac{740.7}{375} \frac{687.6}{16} = 351.1 \ [\text{Nm}]$$

The required torque at the motor shaft during the braking mode is:

$$T_{m,b} = T_r - \frac{(GD^2)_{shaft}}{375} \frac{dn}{dt} = 266.2 - \frac{740.7}{375} \frac{687.6}{10} = 130.4 \ [\text{Nm}]$$

The load diagram (figure E8.6.1)

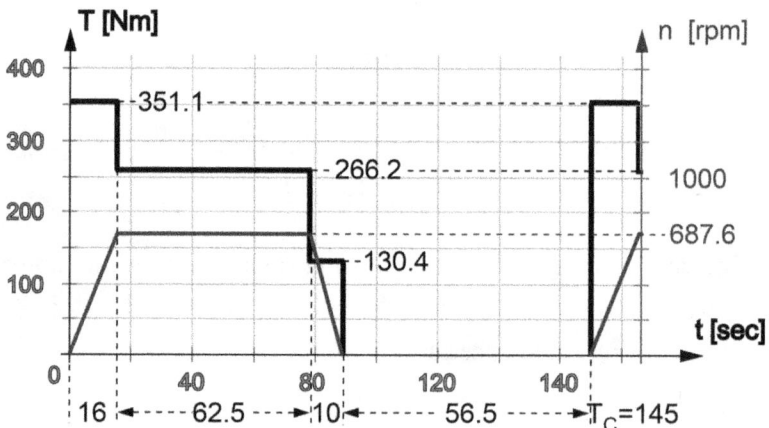

Figure E8.6.1. Load diagram of the industrial rolling machine.

(b) The equivalent torque (equation (8.29)) at the motor shaft, when taking into consideration the fact that the motor is self-cooled (equation (8.59)), is:

$$
\left\{
\begin{aligned}
T_{eq} &= \sqrt{\frac{T_{m,\,s}^2\, t_s + T_w^2 t_w + T_{m,\,b}^2\, t_b}{(t_s + t_b)\alpha + t_w + t_r\beta}} \\
&= \sqrt{\frac{451.1^2 \cdot 16 + 266.2^2 \cdot 62.5 + 130.4^2 \cdot 10}{(16 + 10) \cdot 0.5 + 62.5 + 56.5 \cdot 0.25}} = 270.8 \ [\text{Nm}]
\end{aligned}
\right\}
$$

The equivalent power at the motor shaft is:

$$P_{eq} = T_{eq} \cdot \omega_n = 270.8 \cdot 72 = 19.5 \ [\text{kW}]$$

The calculated equivalent power above, is smaller than the estimated value, $P_{eq} < P_{est}$. This suggests two options:

(1) select a motor that has rated a power between $19.5 < P_n < 23$ kW, or
(2) re-estimate a new rated power for the motor and repeat the dynamic torque calculation to modify the load diagram.

In addition, the chosen motor must be assessed for starting and maximum torques, and for ambient conditions.

8.7 Problems

1. The nameplate of a DC motor states $V_n = 440$ V, $P_n = 22$ kW, $I_n = 56$ A, $n_n = 850$ rpm. The motor operates continuously at an ambient temperature of 35°C and reaches the maximum temperature of 85°C. Assuming a motor loss coefficient of $a = 1$, what would be the motor power capacity at an ambient temperature of 30°C?

2. The nameplate of a 3-ϕ induction motor states $V_n = 208$ V, $P_n = 10$ kW, $n_n = 720$ rpm, $\eta_n = 0.85$, and $(PF)_n = 0.81$. At an ambient temperature of 35°C, the motor operates continuously at full load and reaches a maximum temperature of 80°C. Assuming a motor loss coefficient of $a = 0.8$, what would be its power capacity at an ambient temperature of 60°C?

3. A 3-ϕ induction motor has the following parameters: $P_n = 23$ kW, $I_n = 42.8$ A, $\eta_n = 0.84$, and $R_{sc} = R_1 + R_2' = 0.56$ Ω/ph. The catalogue ambient temperature is $\theta_0^{(cat)} = 35°C$, and the maximum allowed temperature is $\theta_{mx}^{(cat)} = 105°C$. Neglect the magnetizing current of the motor and calculate:
   (a) The power capacity of the motor at an ambient temperature of 65 °C.
   (b) The peak ambient temperature that would prevent loading the motor.

4. A 3-ϕ induction motor has the following parameters: input voltage $V_L = 460$ V line-to-line, rated power $P_n = 24$ kW, and rated slip $S_n = 0.04$. Its simplified equivalent diagram per phase is shown in figure P8.4 where the winding resistance $R_1 = R_2' = 0.2$ Ω/ph, the leakage reactance $X_1 = X_2' = 1.8$ Ω/ph, the core loss resistance $R_a = 139.2$ Ω/ph, and the magnetizing reactance $X_m = 26.5$ Ω/ph (see chapter 5).

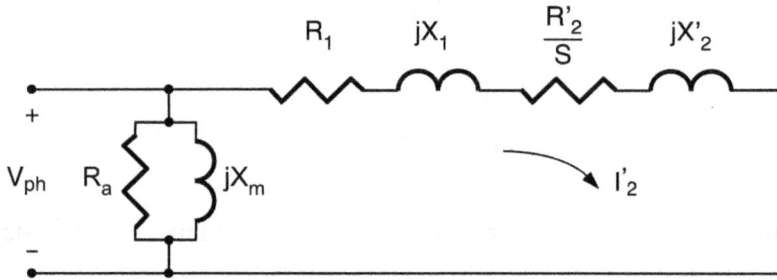

Figure P8.4. Simplified equivalent diagram per phase of an induction motor.

The catalogue (rated) maximum temperature rise of the motor is $\theta_{mx,n} = 85$ °C, and its rated ambient temperature is $\theta_0^{(cat)} = 20$ °C.

The motor has to operate in a room having an ambient temp of 45 °C. Neglect the friction losses, and calculate the de-rated nameplate power of the motor.

5. The load diagram at the shaft of an externally cooled motor is given in figure P8.5, where the cycle time is $T_C = 5.6$ min.

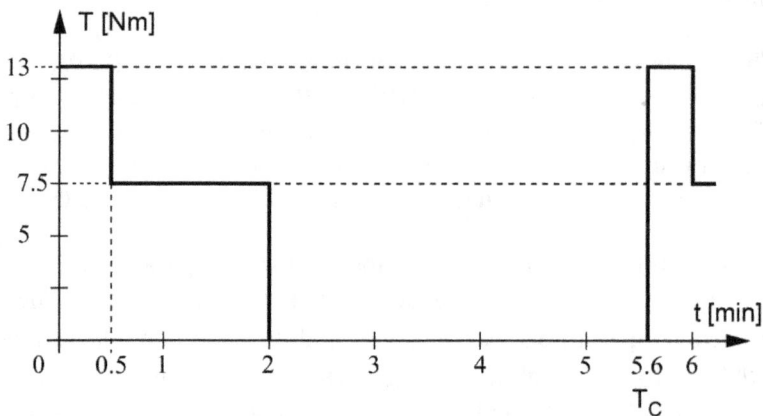

Figure P8.5. Load diagram at the shaft of an externally cooled induction motor.

A catalogue of intermittent operating motors provides the following data:

$$\begin{cases} \text{Motor \#1:} & P_{n1}^{(cat)} = \ \ 0.7 \text{ kW at } \varepsilon_{n1}^{(cat)} = 0.60 \\ \text{Motor \#2:} & P_{n2}^{(cat)} = 0.8 \text{ kW at } \varepsilon_{n2}^{(cat)} = 0.45 \\ \text{Motor \#3:} & P_{n3}^{(cat)} = 0.9 \text{ kW at } \varepsilon_{n3}^{(cat)} = 0.35 \\ \text{Motor \#4:} & P_{n4}^{(cat)} = 1.1 \text{ kW at } \varepsilon_{n4}^{(cat)} = 0.25 \\ \text{Motor \#5:} & P_{n5}^{(cat)} = 1.4 \text{ kW at } \varepsilon_{n5}^{(cat)} = 0.15 \end{cases}$$

The rated speed of all five motors is $n_n = 955$ rpm.

Neglect the constant (iron and friction) losses of the motor, and select the appropriate motor for the above load diagram.

6. An externally cooled induction motor drives a mechanism. The load diagram at the motor shaft is presented in figure P8.6, where the cycle time is $T_C = 11$ min. The rated speed of the motor is $n_n = 1150$ rpm. When the motor operates at rated load, the constant losses present 40% of its total power loss. The drive system is to be operated at an ambient temperature of $-20\ °C$.

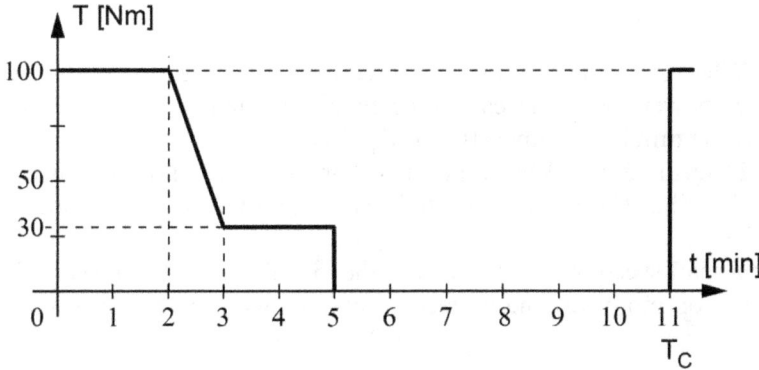

Figure P8.6. Load diagram at the shaft of the induction motor.

A catalogue for intermittent operating motors offers the following data:

$$\begin{cases} \text{Motor \#1:} & P_{n1}^{(cat)} = \ \ 20 \text{ kW at } \varepsilon_{n1}^{(cat)} = 0.2 \\ \text{Motor \#2:} & P_{n2}^{(cat)} = \ \ 20 \text{ kW at } \varepsilon_{n2}^{(cat)} = 0.3 \\ \text{Motor \#3:} & P_{n3}^{(cat)} = \ \ 20 \text{ kW at } \varepsilon_{n3}^{(cat)} = 04 \\ \text{Motor \#4:} & P_{n4}^{(cat)} = \ \ 20 \text{ kW at } \varepsilon_{n4}^{(cat)} = 0.5 \end{cases}$$

The catalogue ambient temperature of each of the four motors is $\theta_0^{(cat)} = 20°C$, and their maximum temperature is $\theta_{mx}^{(cat)} = 110°C$.

Select the appropriate motor for the above load diagram.

7. A load diagram at the shaft of a self-ventilated induction motor is given in figure P8.7. The rated shaft speed is $n_n = 1140$ rpm, and the cycle time is $T_C = 90$ min. The motor operates in a refrigerated area where the ambient temperature is $-10\,°C$. The start/brake cooling coefficient is $\alpha = 0.6$, and the rest coefficient is $\beta = 0.25$.

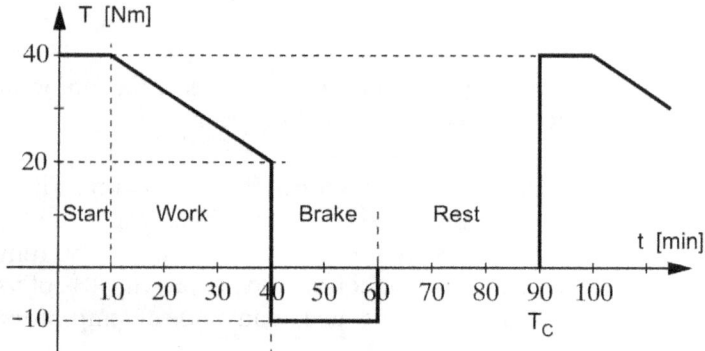

Figure P8.7. Load diagram at the self-ventilated motor.

When the motor operates at rated load, the core losses present 40% of its total power loss. The catalogue maximum temperature is $\theta_{mx}^{(cat)} = 180°C$, and the ambient temperature is $\theta_0^{(cat)} = 20°C$.

Determine: (a) What type of motor catalogue should be used?
(b) What should be the rated (catalogue) power of the motor.

8. At a shaft speed of $n = 1750$ rpm, the load diagram, torque versus time, of an induction motor driving a mechanism is shown in figure P8.8 where the cycle

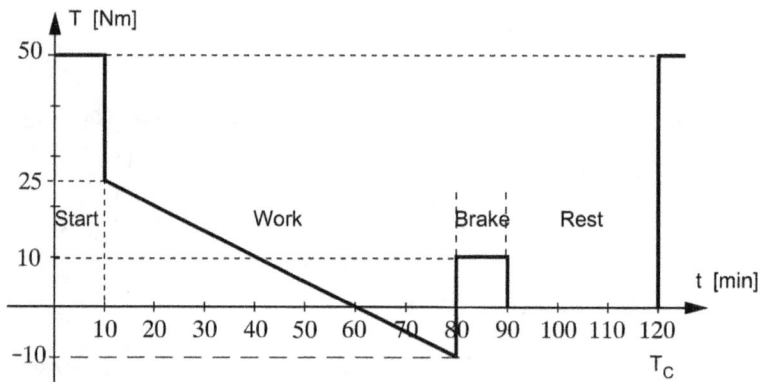

Figure P8.8. Load diagram—torque versus time.

time is $T_C = 120$ min. The motor is a self-cooling machine where the start/brake cooling coefficient is $\alpha = 0.5$, and the rest coefficient is $\beta = 0.25$.

Calculate the required power at the motor shaft.

9. An electromechanical drive system employs a mechanism that operates at an intermittent periodic mode where the cycle time is $T_C = 10$ min. The mechanism is driven by a self-ventilated induction motor at a speed of $n_n = 1710$ rpm. The motor loss coefficient is $a = 1.1$.

The load diagram, as seen at the motor shaft, is given in figure P8.9. The complete drive system should operate in a refrigerated environment at $-15\,°C$.

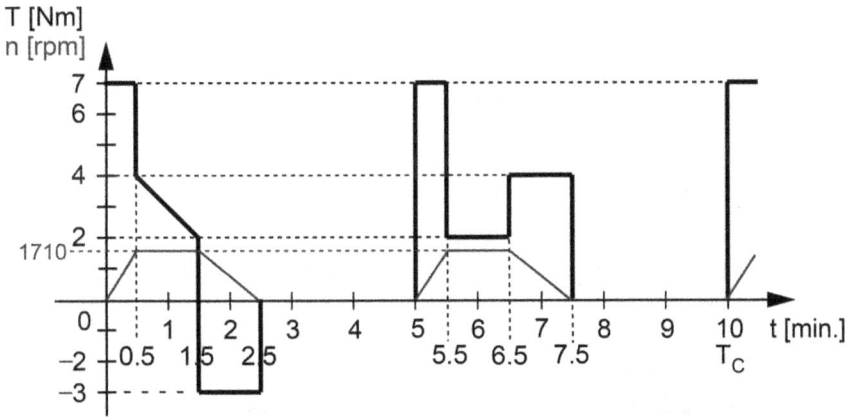

Figure P8.9. Load diagram at the motor shaft.

A catalogue for intermittent operating motors offers the following data:

$$
\begin{cases}
\text{Motor \#1:} & P_{n1}^{(cat)} = 1150 \text{ W at } \varepsilon_{n1}^{(cat)} = 0.3 \\
\text{Motor \#2:} & P_{n2}^{(cat)} = 1150 \text{ W at } \varepsilon_{n2}^{(cat)} = 0.4 \\
\text{Motor \#3:} & P_{n3}^{(cat)} = 1150 \text{ W at } \varepsilon_{n3}^{(cat)} = 0.5 \\
\text{Motor \#4:} & P_{n4}^{(cat)} = 1150 \text{ W at } \varepsilon_{n4}^{(cat)} = 0.6
\end{cases}
$$

The catalogue ambient temperature of each of the four motors is $\theta_0^{(cat)} = 20°C$, and their maximum temperature is $\theta_{mx}^{(cat)} = 105°C$.

Select the appropriate motor that would address the given load diagram.

10. A steel-made moving platform is driven by a self-ventilated AC motor. The stationary motor has a gearbox and a cogwheel on its shaft (motor and gearbox are located behind the cogwheel, and are not shown in the figure P8.10). The platform moves to the right, stops for a short time, and then moves to the left, stops again, in a cyclic operation, where the cycle time is $T_C = 9$ s.

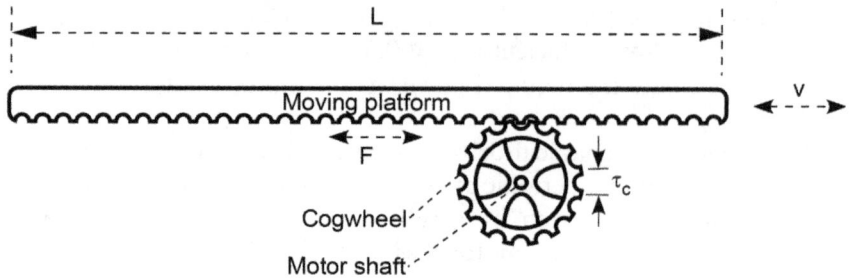

Figure P8.10. A cyclic operation moving platform.

Given:

The platform length is $L = 3$ m.

The platform velocity, in either direction, is $v = 1$ m s^{-1}.

The platform acceleration rate is $a_s = 4$ m s^{-1}.

The platform deceleration applied by mechanical friction is $a_b = 5$ m s^{-1}.

The force required to move the platform, in either direction, is $F = 2125$ N.

The weight of the moving platform is $G_P = 350$ kg$_f$.

The moment of inertia of the cogwheel is $(GD^2)_c = 38$ Nm^2.

The number of teeth of the cogwheel is $Z_c = 16$, and its tooth pitch is $\tau_c = 25$ mm.

The moment of inertia of the motor can be assumed: $(GD^2)_m = 0.28$ Nm^2.

The moment of inertia of the gearbox is $(GD^2)_g = 0.2$ Nm^2.

The gearbox ratio is $i = 11.5$.

The efficiency of the gearbox is $\eta_g = 0.95$.

The efficiency of the cogwheel and the moving platform is $\eta_c = 0.85$.

Calculate:

(a) The duty cycle of the electromechanical drive ε.

(b) The load diagram of the drive system (torque versus time).

(c) The required rated power of the motor.

References

[1] Mohan N 2001 *Electric Drives: An Integrative Approach* (MNPERE Publisher)

[2] Dubey G K 2001 *Fundamentals of Electrical Drives* (Alpha Science International Ltd)

[3] Chilikin M 1976 *Electric Drive* (Moscow: MIR Publishers)

[4] ANSI/NEMA MG 1-2016 (revised 2018), Section IV

[5] Zwillinger D 2012 *Standard Mathematical Tables and Formulae* 32nd edn (Boca Raton, FL: CRC Press)

www.ingramcontent.com/pod-product-compliance
Lightning Source LLC
Chambersburg PA
CBHW080543220326

41599CB00032B/6349